Springer Theses

Recognizing Outstanding Ph.D. Research

Aims and Scope

The series "Springer Theses" brings together a selection of the very best Ph.D. theses from around the world and across the physical sciences. Nominated and endorsed by two recognized specialists, each published volume has been selected for its scientific excellence and the high impact of its contents for the pertinent field of research. For greater accessibility to non-specialists, the published versions include an extended introduction, as well as a foreword by the student's supervisor explaining the special relevance of the work for the field. As a whole, the series will provide a valuable resource both for newcomers to the research fields described, and for other scientists seeking detailed background information on special questions. Finally, it provides an accredited documentation of the valuable contributions made by today's younger generation of scientists.

Theses are accepted into the series by invited nomination only and must fulfill all of the following criteria

- They must be written in good English.
- The topic should fall within the confines of Chemistry, Physics, Earth Sciences, Engineering and related interdisciplinary fields such as Materials, Nanoscience, Chemical Engineering, Complex Systems and Biophysics.
- The work reported in the thesis must represent a significant scientific advance.
- If the thesis includes previously published material, permission to reproduce this must be gained from the respective copyright holder.
- They must have been examined and passed during the 12 months prior to nomination.
- Each thesis should include a foreword by the supervisor outlining the significance of its content.
- The theses should have a clearly defined structure including an introduction accessible to scientists not expert in that particular field.

More information about this series at http://www.springer.com/series/8790

Nils Olaf Bernd Lüttschwager

Raman Spectroscopy of Conformational Rearrangements at Low Temperatures

Folding and Stretching of Alkanes in Supersonic Jets

Doctoral Thesis accepted by
the Georg-August-University Göttingen, Germany

Springer

Author
Dr. Nils Olaf Bernd Lüttschwager
Institute of Physical Chemistry
Georg-August-University Göttingen
Göttingen
Germany

Supervisor
Prof. Dr. Martin A. Suhm
Institute of Physical Chemistry
Georg-August-University Göttingen
Göttingen
Germany

ISSN 2190-5053 ISSN 2190-5061 (electronic)
ISBN 978-3-319-08565-4 ISBN 978-3-319-08566-1 (eBook)
DOI 10.1007/978-3-319-08566-1

Library of Congress Control Number: 2014944327

Springer Cham Heidelberg New York Dordrecht London

Printed on acid-free paper

Springer is part of Springer Science+Business Media (www.springer.com)

Parts of this thesis have been be published in the following journal articles and will be cited by their Roman numerals:

I N. O. B. Lüttschwager, T. N. Wassermann, R. A. Mata and M. A. Suhm, *The Last Globally Stable Extended Alkane*, *Angewandte Chemie International Edition*, 2013, **52**, 463–466, DOI: 10.1002/anie.201202894;

N. O. B. Lüttschwager, T. N. Wassermann, R. A. Mata and M. A. Suhm, *Das letzte Alkan mit gestreckter Grundzustandskonformation*, *Angewandte Chemie*, 2013, **125**, 482–485, DOI: 10.1002/ange.201202894.

II N. O. B. Lüttschwager and M. A. Suhm, *Stretching and folding of 2-nanometer hydrocarbon rods, Soft Matter*, 2014, **10**, 4885–4901, DOI: 10.1039/c4sm00508b.

III P. Drawe, N. O. B. Lüttschwager and M. A. Suhm, *The elastic modulus of isolated polytetrafluoroethylene filaments*, *ScienceOpen Research*, **2014**, DOI: 10.14293/a2199-1006.01.sor-matsci.ka0j6.v1. Published under creative commons license CC BY 4.0: `http://creativecommons.org/licenses/by/4.0`.

Supervisor's Foreword

If one sets out to design a chain polymer, there are at least three things to keep an eye on: the longitudinal stiffness of the individual molecular chains, their conformational flexibility, and the mutual stickiness of the chains. Acting together, these molecular constraints shape key macroscopic properties of the bulk polymer. Is any experimental technique able to quantitatively determine the balance between all three aspects for the most important synthetic polymer on earth—polyethylene—without ever studying polyethylene itself? The answer is "yes"—it is Raman spectroscopy of linear alkanes in adiabatic gas expansions. In your hands, you hold the first detailed description on how to get all that.

Nils Lüttschwager has worked toward this goal in a systematic way for more than three years. He has improved a spontaneous Raman scattering setup capable of probing supersonic jet expansions to the point where the spectroscopy of strictly isolated long chain alkanes with up to 21 segments becomes possible. By careful optimization and characterization in the entire fundamental Stokes excitation window, he has collected unambiguous and redundant spectroscopic evidence that van-der-Waals-driven folding of alkanes occurs beyond 17 carbon atoms. Even longer stretched alkane conformations can be kinetically stabilized and the evolution of their accordion mode with chain length allows for a parameter-free extrapolation of the elastic modulus for an infinite polyethylene chain. The Raman spectra can be modeled with superb fidelity in terms of an effective conformational temperature of about 100 K for single gauche isomerizations, whereas the full hairpin turns are frozen in at a higher temperature. Conformational cooling is essential to resolve the spectra and a vacuum environment is essential to avoid interactions between the molecules, which would prevent folding, and affect the force required for longitudinal stretching. On the theoretical side, the thesis features quantum and statistical mechanical simulations capturing the key aspects.

For the more bulky polytetrafluoroethylene, Nils Lüttschwager shows by an analogous experimental strategy that the same force is required for chain elongation as for alkanes, whereas the cross-section-dependent elastic modulus is naturally smaller. Furthermore, the investigated perfluoroalkanes show a pronounced propensity for stretched-out or actually slightly helical conformations, which is related to the exceptional frictional properties of this material.

No molecular model of long alkyl chains (be it in polymers or biological membranes) should be considered accurate if it cannot reproduce the experimental benchmark values derived in this work within appropriate error bars. The thesis provides a monographic, in-depth description of the spectroscopic work and its analysis in the context of conformational, valence, and London dispersion forces controlling linear alkanes. Its key findings may even qualify for being mentioned in undergraduate textbooks of organic chemistry and materials physics. Enjoy this unique monograph on materials science in the gas phase!

Göttingen, March 2014 Prof. Dr. Martin A. Suhm

Acknowledgments

I want to express my gratitude to all people who contributed to the success of this work. First, of course, I thank my supervisor and the first examiner of this work, Professor Martin Suhm, for the exciting task, the freedom, and his tireless support and enthusiasm. Furthermore, I am thankful to Professor Jörg Schroeder, the co-examiner of this work, and all members of the examination board who surveyed this thesis. I like to thank Dr. Michael Hippler, senior lecturer at the University of Sheffield, for giving me the opportunity to dip into optical cavity enhancement, which will be applied by us in Göttingen in the near future. A special thanks goes to the Deutsche Forschungsgemeinschaft for the financial support for this work.

Beside my supervisors, I had great support from other Ph.D. students and postdocs from the Suhm group, first and foremost from Dr. Tobias Wassermann. Tobias supported me right from the beginning of my diploma studies, investigated alkanes, and left me an excellent Raman spectrometer for my own alkane research. Another researcher at the "curry-jet" was Dr. Zhifeng Xue, who was always willing to help me with all the troubles I had with the apparatus.

None of the spectra presented in this thesis would have been measured without the expertise of our various workshops. Representative for the whole staff I thank Volker Meyer, head of the mechanical workshop, Andreas Knorr, head of the electronic workshop, and Annika von Roden, head of the glass workshop, for their exceptional work. Werner "Milo" Noack knows about all the treasures hidden somewhere in the institute that want to be rediscovered. Milo helped me a great deal with equipment and his knowledge about manufacturers and I thank him for that. For support regarding paperwork, I thank Petra Lawecki and Dr. Markus Hold.

My sincere thanks go to the whole Suhm group for the marvelous time, the mutual support, pleasant group excursions, and the daily table soccer match, of course. Special thanks go to Franz Kollipost with whom I visited several conferences and who contributed to this work in numerous discussions. Also, I want to thank Patrick Drawe for measuring perfluorinated alkanes, Fabian Dietrich for helping me with the upgrade of the Raman spectrometer, and Sebastian Bocklitz for helping me with the measurements of the C–C stretching region.

Just as the Suhm group, my fellow students and friends are the reason that I always felt comfortable and welcome in Göttingen and at the university: Dr. Tim Hungerland, Kai Kalz, Dr. Johannes "Jones" Kaschel und Dr. Markus Leibeling. I wish you all the best for the future, and I wish us that we stay in touch! To my family and beloved ones, my mother Gaby, Albrecht, my sisters Nadja and Nora, and Alex: Thank you for your support in an exciting time!

Göttingen, March 2014 Nils Lüttschwager

Contents

Chapter 1
Introduction

> *My interest in science is to simply find out about the world, and the more I find out the better it is. I like to find out.*
>
> Richard Feynman

A large part of chemistry deals with how atoms bind and how to build any desired, but yet physically possible molecule. This addresses mainly the connectivity of chemically bound atoms, the constitution, which largely determines how a particular molecule reacts, as well as the configuration, which describes the rigid arrangement of atoms in space and might be just as important for the molecular functionality. Still, constitution and configuration do not fully answer the question how a flexible molecule actually "looks like", that is, how the atoms are arranged around single bonds. The exact arrangement is covered by the conformation, which takes rotations about single bonds into account. Small and simple molecules may have just one or a few stable conformers, but the conformational flexibility increases quickly with the number of single bonds. For large and complex chain molecules, such as those making up living organisms, there exist vast numbers of stable conformations, and yet the occupation of the correct conformation is crucial to their functionality. A popular example are proteins [1], which can only function correctly if they are properly 'folded', that is, if they occupy the appropriate conformation. Misfolding leads to improper functioning or may even lead to disease [2]. Therefore, understanding why a particular conformation is preferred and what factors influence conformational preferences is vital when working with complex chemical systems. Among the interactions [3, 4] which determine the stability of molecular conformations are the ubiquitous attractive dispersion forces (London forces), caused by instantaneous dipole (and higher order) moments due to fluctuation in electron density, as well as steric repulsion of electron pairs at short atom distances. The main goal of this thesis is to quantify the interplay between dispersion stabilization and short-range repulsion by studying preferred conformations of unbranched n-alkanes by means of vibrational spectroscopy.

Alkanes are an obvious choice for such a study, because they lack substantially polar groups which would superimpose permanent electrostatic interactions or functional groups which would allow the formation of hydrogen bonds. Rotation

N.O.B. Lüttschwager, *Raman Spectroscopy of Conformational Rearrangements at Low Temperatures*, Springer Theses, DOI 10.1007/978-3-319-08566-1_1

about single carbon-carbon bonds in alkane chains involves relatively low energy barriers and conformers can be converted quickly into one another [5]. Alkanes with not too long chain lengths prefer a fully stretched conformation where all carbon-carbon torsional angles assume a value of 180°. In this *all-trans* conformation, the steric repulsion of adjacent chain segments is minimal. Another stable conformation exists at a torsional angle of about ±65°, which is termed synclinal or gauche conformation. Twisting a single carbon-carbon bond of an all-trans alkane from the trans to the gauche conformation raises the energy of the molecule (by about 2–3 kJ mol^{-1} per gauche [6–9]), but several of such twists allow to align two segments of the chain in a way that profits from stabilizing dispersion interaction. The dispersion stabilization increases with the length of aligned chain segments, while the energy cost to fold the molecule stays about the same, such that after a certain *critical chain length* dispersion stabilization wins and the folded conformer is more stable than the all-trans conformer. Thus, the appropriate question to ask in this context is: "What is the longest unbranched alkane with a linear global minimum conformation?" [10]. This is the central question which will be addressed here.

Why is this interesting? The general importance of intra- (and inter-) molecular interactions was indicated above, but more specifically, the folding or "self-solvation" of alkanes is an excellent and sensitive benchmark for computational chemistry [10–13]. The importance of computational chemistry was acknowledged by the 2013 Noble Prize in Chemistry, awarded to M. Karplus, M. Levitt, and A. Warshel for their contribution to the field of molecular modeling. [1] Testing such models by rigorous experimental benchmarks is necessary to establish their validity and promote their development. To predict the correct critical chain length, a computational method needs to correctly describe the torsion potentials which determine how much energy the folding of an alkane chain costs, as well as the dispersion forces which stabilize a folded conformer.

A tightly folded conformer results from the conformational sequence gauche-gauche-trans-gauche-gauche (ggtgg) where all gauche conformations have the same sense of rotation. Since the uniform helicity is absolutely necessary to align two chain segments by this sequence, this may be termed an intramolecular case of chirality synchronization [14, 15]. One of the resulting enantiomeric 'hairpin' conformers is depicted in Fig. 1.1, together with a qualitative plot which illustrates how the hairpin conformer becomes more stable with increasing chain length and finally surpasses the stability of the all-trans conformer beyond the critical chain length (n_c). This hairpin/all-trans competition depends critically on the molecular environment. Alkanes in water, for example, minimize their surface area exposed to the solvent by folding [16, 17], such that the critical chain length in this environment decreases [10]. On the other hand, more favorable dispersion interactions with an unpolar solvent or neighboring alkanes in pure solids withdraw the driving force behind self-solvation, postponing its occurrence to much larger chain lengths [10, 18, 19]. In order to find

[1] http://www.nobelprize.org/nobel_prizes/chemistry/laureates/2013/advanced-chemistryprize2013.pdf (15 October 2013).

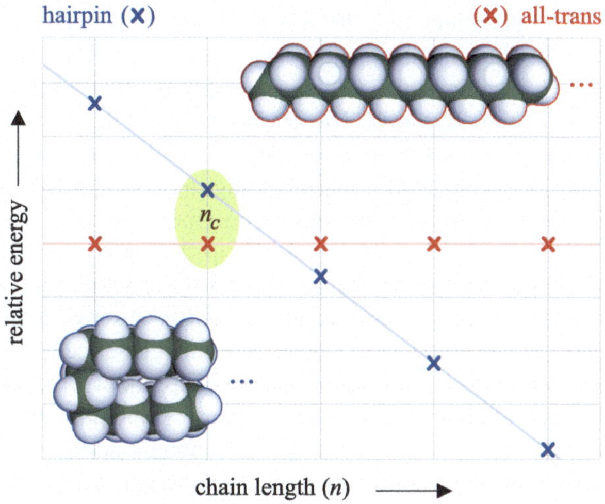

Fig. 1.1 Qualitative plot of the energy of a folded hairpin conformer relative to the energy of the all-trans conformer. n_c marks the critical chain length beyond that the all-trans conformer is not the global minimum structure anymore. Reference II—Reproduced by permission of The Royal Society of Chemistry.

the "true" critical chain length of alkanes free of intermolecular interactions, one has to employ studies in the gas phase.

Current calculations of interaction-free alkanes find a critical chain length in the range $n_c = 15$–21 [11–13]. The vapor pressure of alkanes in this size range is very low at room temperature [20]. A gas phase study would thus imply substantial heating, but hot alkanes are ordered rather randomly and low-energy conformations such as the all-trans or hairpin conformer are not sufficiently populated to be analyzed. In this work, alkanes were thus cooled in supersonic jet expansions [21, 22] to shift the conformational population from random gauche conformers to all-trans and hairpin conformers. By first heating and subsequent expansion it was possible to investigate cold and isolated alkanes in the relevant size range up to a chain length of 21 carbon atoms for the first time. To determine the critical chain length, jet-cooled alkanes were investigated by vibrational Raman spectroscopy—a routine tool to investigate structural preferences of alkanes and alkyl-systems [23–25]. Although one can resort to numerous spectroscopic investigations of liquid and solid alkanes, especially from T. Shimanouchi, R.G. Snyder and G. Zerbi, Raman spectroscopic studies of alkanes in the gas phase are rare and rather restricted to short chains [9, 26, 27], not least because the preparation of cold gaseous alkanes is technically challenging.

Especially useful to analyze Raman spectra are low-frequency frame vibrations, like the strongly Raman active "accordion" or longitudinal acoustic modes (LAMs) of all-trans segments [28], which have wavenumbers related to the length of the vibrating segment [29]. They will be of particular interest to this work. Observation of accordion vibrations allows furthermore to derive a value for the ultimate elastic modulus of polyethylene [27, 30, 31], the polymer counterpart of n-alkanes. With the same aim, perfluorinated alkanes were investigated. They are an interesting comparative case to n-alkanes which relates to the technically useful polymer polytetrafluoroethylene (PTFE or Teflon®). The determination of the elastic modulus links this work to polymer physics, but spectroscopy of small oligomers in general [32], and conformational preferences like sharp folding in particular [33–35] are also of interest to deduce the microscopic structure of polymers [36, 37].

The outline of this thesis is as follows. In Chap. 2, Raman spectroscopy will be introduced briefly, focusing on how Raman intensities of vibrational transitions are calculated. The supersonic expansion technique will be discussed there as well. In Chap. 3, the experimental setup used to measure Raman jet-spectra will be outlined, the so-called curry-jet.[2] This includes a discussion of upgrades which were necessary to allow stable heating of alkanes prior to the expansion and how different measurement conditions influence n-alkane jet-spectra. Chapter 4 addresses n-alkanes and provides information on stable conformations, characteristic vibrations, how spectra are simulated and assigned, and ultimately how the critical chain length is derived from the experimental observations. Chapter 5 deals with perfluorinated alkanes, including Raman jet-spectra of low-frequency vibrations and a discussion regarding self-solvation. Chapter 6 closes the discussion part with the determination of elastic moduli of polyethylene and polytetrafluoroethylene. A summary of the findings of this dissertation is provided in Chap. 7.

References

1. C.M. Dobson, Protein folding and misfolding. Nature **426**, 884–890 (2003)
2. E. Reynaud, Protein misfolding and degenerative diseases. Nature Educ. **3**, 28 (2010)
3. S.K. Burley, G.A. Petsko, in *Advances in Protein Chemistry*, vol. 39, ed. by C. Anfinsen, J.T. Edsall, F.M. Richards, D.S. Eisenberg, (Academic Press, New York, 1988), pp. 125–189
4. A. Stone, *The Theory of Intermolecular Forces*, 2nd edn. (Oxford University Press, Oxford, 2013)
5. Y. Mo, A critical analysis on the rotation barriers in butane. J. Org. Chem. **75**, 2733–2736 (2010)
6. W.A. Herrebout, B.J. van der Veken, A. Wang, J.R. Durig, Enthalpy difference between conformers of n-butane and the potential function governing conformational interchange. J. Phys. Chem. **99**, 578–585 (1995)
7. J.B. Klauda, B.R. Brooks, A.D. MacKerell, R.M. Venable, R.W. Pastor, An ab initio study on the torsional surface of alkanes and its effect on molecular simulations of alkanes and a DPPC bilayer. J. Phys. Chem. B **109**, 5300–5311 (2005)
8. D. Gruzman, A. Karton, J.M.L. Martin, Performance of ab initio and density functional methods for conformational equilibria of C_nH_{2n+2} alkane isomers (n = 4–8). J. Phys. Chem. A **113**, 11974–11983 (2009)
9. R.M. Balabin, Enthalpy difference between conformations of normal alkanes: Raman spectroscopy study of n-pentane and n-butane. J. Phys. Chem. A **113**, 1012–1019 (2009)

[2] The name of the setup is a culinary homage to the Indian discoverer of the Raman-effect, Chandrasekhara Venkata Raman, and stands for classical unrestricted Raman spectroscopy in a jet.

10. J.M. Goodman, What is the longest unbranched alkane with a linear global minimum conformation? J. Chem. Inf. Comput. Sci. **37**, 876–878 (1997)
11. L.L. Thomas, T.J. Christakis, W.L. Jorgensen, Conformation of alkanes in the gas phase and pure liquides. J. Phys. Chem. B **110**, 21198–21204 (2006)
12. S. Grimme, J. Antony, T. Schwabe, C. Mück-Lichtenfeld, Density functional theory with dispersion corrections for supramolecular structures, aggregates, and complexes of (bio)organic molecules. Org. Biomol. Chem. **5**, 741–758 (2007)
13. T. Schwabe, S. Grimme, Double-hybrid density functionals with long-range dispersion corrections: higher accuracy and extended applicability. Phys. Chem. Chem. Phys. **9**, 3397–3406 (2007)
14. A. Zehnacker, M.A. Suhm, Chirality recognition between neutral molecules in the gas phase. Angew. Chem. Int. Ed. **47**, 6970–6992 (2008)
15. A. Zehnacker, M.A. Suhm, Chiralitätserkennung zwischen neutralen Molekülen in der Gasphase. Angew. Chem. **120**, 7076–7100 (2008)
16. A. Wallqvist, D.G. Covell, Free-Energy cost of bending n-dodecane in aqueous solution, influence of the hydrophobic effect and solvent exposed area. J. Phys. Chem. **99**, 13118–13125 (1995)
17. S. Chakrabarty, B. Bagchi, Self-Organization of n-alkane chains in water: length dependent crossover from helix and toroid to molten globule. J. Phys. Chem. B **113**, 8446–8448 (2009)
18. K.S. Lee, G. Wegner, Linear and cyclic alkanes (C_nH_{2n+2}, C_nH_{2n}) with $n > 100$, synthesis and evidence for chain-folding. Die Makromol. Chem. Rapid Commun. **6**, 203–208 (1985)
19. G. Ungar, J. Stejny, A. Keller, I. Bidd, M.C. Whiting, The crystallization of ultralong normal paraffins: the onset of chain folding. Science **229**, 386–389 (1985)
20. *CRC Handbook of Chemistry and Physics*, ed. by D.R. Lide, 82nd edn, (CRC Press, Boca Raton, 2001)
21. D.H. Levy, The spectroscopy of very cold gases. Science **214**, 263–269 (1981)
22. M. Herman, R. Georges, M. Hepp, D. Hurtmans, High resolution Fourier transform spectroscopy of jet-cooled molecules. Int. Rev. Phys. Chem. **19**, 277–325 (2000)
23. R.G. Snyder, On Raman evidence for conformational order in liquid n-alkanes. J. Chem. Phys. **76**, 3342–3343 (1982)
24. C.J. Orendorff, M.W. Ducey Jr, J.E. Pemberton, Quantitative correlation of Raman spectral indicators in determining conformational order in alkyl chains. J. Phys. Chem. A **106**, 6991–6998 (2002)
25. L. Brambilla, G. Zerbi, Local order in liquid n-alkanes: evidence from Raman spectroscopic study. Macromolecules **38**, 3327–3333 (2005)
26. N.A. Atamas, A.M. Yaremko, T. Seeger, A. Leipertz, A. Bienko, Z. Latajka, H. Ratajczak, A.J. Barnes, A study of the Raman spectra of alkanes in the Fermi-resonance region. J. Mol. Struct. **708**, 189–195 (2004)
27. T.N. Wassermann, J. Thelemann, P. Zielke, M.A. Suhm, The stiffness of a fully stretched polyethylene chain: a Raman jet spectroscopy extrapolation. J. Chem. Phys. **131**, 161108 (2009)
28. R.G. Snyder, The structure of chain molecules in the liquid state: low-frequency Raman spectra of n-alkanes and perfluoro-n-alkanes. J. Chem. Phys. **76**, 3921–3927 (1982)
29. R.F. Schaufele, T. Shimanouchi, Longitudinal acoustical vibrations of finite polymethylene chains. J. Chem. Phys. **47**, 3605–3610 (1967)
30. R.G. Snyder, H.L. Strauss, R. Alamo, L. Mandelkern, Chain-length dependence of interlayer interaction in crystalline n-alkanes from Raman longitudinal acoustic mode measurements. J. Chem. Phys. **100**, 5422–5431 (1994)
31. G.D. Barrera, S.F. Parker, A.J. Ramirez-Cuesta, P.C.H. Mitchell, The vibrational spectrum and ultimate modulus of polyethylene. Macromolecules **39**, 2683–2690 (2006)
32. *Modern Polymer Spectroscopy*, ed. by G. Zerbi, (Wiley-VCH, Weinheim, 1999)
33. G. Zerbi, M. Gussoni, Defect modes for (200), GGTGG, tight fold re-entry in polyethylene single crystals. Polymer **21**, 1129–1134 (1980)

34. S. Wolf, C. Schmid, P.C. Hägele, Vibrational analysis of the tight (110) fold in polyethylene. Polymer **31**, 1222–1227 (1990)
35. G. Ungar, X.B. Zeng, S.J. Spells, Non-integer and mixed integer forms in long *n*-alkanes observed by real-time LAM spectroscopy and SAXS. Polymer **41**, 8775–8780 (2000)
36. R. Eckel, M. Buback, G.R. Strobl, Untersuchung der druckinduzierten Kristallisation von Polyäthylen mit Hilfe einer neuen Raman-Hochdruckzelle. Colloid Polym. Sci. **259**, 326–334 (1981)
37. G.R. Strobl, W. Hagedorn, Raman spectroscopic method for determining the crystallinity of polyethylene. J. Polym Sci. Polym. Phys. Ed. **16**, 1181–1193 (1978)

Chapter 2
Background

This chapter is intended to give background information on techniques employed in the presented studies. The goal is not to describe these techniques in great detail, but to point out aspects which are of particular importance to this work. The central tool is vibrational Raman spectroscopy, which is used to study cold molecules seeded in supersonic jet expansions. Vibrational Raman spectra are interpreted aided by quantum chemical calculations in the harmonic approximation. Raman spectroscopy is discussed in detail in Refs. [1–3] and many others. Information on supersonic jet-expansions can be found in Refs. [4–8]. An introduction to quantum chemistry is provided by the textbooks [9] and [10], for example.

The next section will address the basic principles of Raman spectroscopy and how Raman intensities are calculated. In Sect. 2.2, the supersonic expansion technique will be introduced briefly. Since supersonic expansions prepared with the curry-jet were systematically characterized previously [11], but not in this work, relations which describe the properties of a supersonic expansion are largely omitted and the discussion is kept rather on a qualitative level. No attempt is made to introduce the wide and rich field of quantum chemistry, but quantum chemical predictions are used frequently and details will be given occasionally.

2.1 Raman Spectroscopy and Scattering Cross-Section

Raman spectroscopy takes advantage of the Raman effect, which can be pictured as photons being inelastically scattered at molecules. The Raman effect is named after its discoverer, Chandrasekhara Venkata Raman [12] who was awarded the Nobel Prize in Physics for its discovery in 1930.[1] Raman scattering is termed inelastic because energy is transfered between the photon and the molecule, causing the photon to change its frequency and the molecule to change its state. The frequency change or

[1] http://www.nobelprize.org/nobel_prizes/physics/laureates/1930/raman-lecture.pdf (21 September 2013).

N.O.B. Lüttschwager, *Raman Spectroscopy of Conformational Rearrangements at Low Temperatures*, Springer Theses, DOI 10.1007/978-3-319-08566-1_2

Raman shift of the scattered light is probed in Raman spectroscopy, with the aim to draw conclusions regarding the molecular states. This work uses Raman spectroscopy to determine changes in vibrational states which allow to deduce the structure or conformation of the investigated substances. More specifically, all spectra presented here are measurements of Stokes scattering, where the photon loses energy to the molecule which is excited to a higher vibrational state. The opposite case where energy is transfered from the molecule in an excited state to the photon, anti-Stokes scattering, is usually much weaker (the population of excited vibrational states is often rather low) and was not considered.

The Raman effect is extremely weak. In a Raman experiment, the vast majority of photons is scattered elastically, which gives rise to a strong signal at the excitation wavelength, the Rayleigh-line. Only one out of 10^6–10^8 photons is scattered inelastically [3] and contributes to Raman signals, which is the reason why this work made use of a powerful excitation laser and sensitive CCD detector.

In order to measure the Raman shift precisely, the employed excitation radiation should be sufficiently monochromatic. In principle, the excitation wavelength (λ) is flexible, and higher frequencies or shorter wavelengths increase the intensity (I) of Raman scattering dramatically ($I \propto \lambda^{-4}$). Yet, much shorter wavelengths than $\lambda \approx 400$ nm (blue-violet light) will often induce electronic transitions and subsequent re-emission due to fluorescence or phosphorescence, which easily outshines Raman scattering. The laser used in this work emits light with a wavelength $\lambda = 532$ nm which is a good compromise and frequently used in Raman experiments.

Scattering Cross-Section

The weakness of Raman scattering is reflected in a very small interaction cross-section or *scattering cross-section*, many orders of magnitude smaller than interaction cross-sections in infrared experiments, where the absorption of photons can be probed to study vibrational excitation. This disadvantage is counterbalanced by the fact that Raman scattering, commonly in the region of visible light, can be detected more efficiently than infrared radiation, such that not the detection sensitivity is determining what technique is applied, but the particular vibrations which are of interest. Raman spectroscopy is the appropriate choice for this work because alkanes and perfluoroalkanes exhibit characteristic vibrations which are strong Raman scatterers but weak infrared absorbers.

The intensity (I) of light scattered perpendicular to the incident beam is proportional to the *differential*[2] scattering cross-section (σ'):

$$I = I_0 \cdot c \cdot \sigma' \cdot a. \tag{2.1}$$

[2] The differential scattering cross-section is a measure of how much scattered light passes through a unit area perpendicular to the incident beam. The total scattering cross-section, on the other hand, is a measure of how much light is removed from the incident beam.

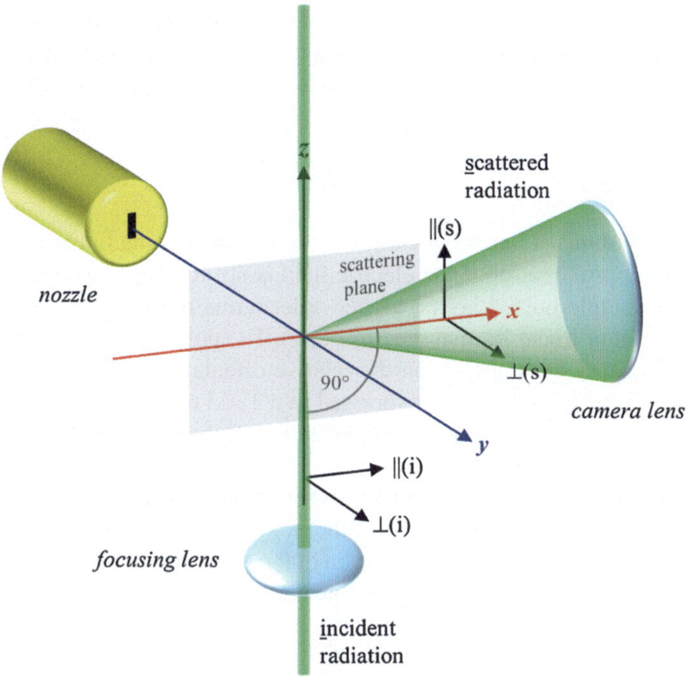

Fig. 2.1 Scattering geometry of the curry-jet setup in an appropriate coordinate system. *Black arrows* indicate the polarization of incident and scattered light

In the following, the specification 'differential' will be dropped and the term scattering cross-section will always refer to the differential scattering cross-section. The scattering cross-section is the intrinsic molecular property which limits the detectability of a particular transition. For spontaneous Raman scattering, the intensity of Raman scattered light depends furthermore linearly on the concentration (c) of the examined substance, and the intensity of the incident radiation (I_0), both of which are controllable to some extend. The constant a invokes experimental factors, such as the probed volume and the solid angle at which scattered light is detected (see Fig. 2.1). Knowing these quantities allows to calculate the absolute intensity of light Raman-scattered by a particular transition, but not all scattered photons are actually detected. In fact, the sensitivity of the curry-jet depends on the wavelength and polarization of the scattered light, which needs to be accounted for when calculations are to be compared to measurements. In general, the concentration of the substance in the probed volume is not known and only a relative comparison is made which can be restricted to scattering cross-sections.

To explain the polarization and orientation dependence of Raman scattering, it is helpful to switch from the particle description to the wave description of light. When a molecule is exposed to electromagnetic radiation, the electric field will distort its electron density and induce an oscillating dipole moment ($\vec{\mu}$) which itself emits radiation at the incident frequency. This is the main[3] source of Rayleigh scattering. The magnitude of the oscillating dipole moment determines the intensity of the emitted radiation ($I \propto |\vec{\mu}|^2$) and depends on how easily the electron density may be distorted. This property is given by the polarizability ($\underline{\alpha}$) of the molecule:

$$\vec{\mu}(t) = \underline{\alpha}\vec{E}(t). \tag{2.2}$$

This equation assumes a linear dependence, which captures the linear Raman effect and is a good approximation if the electric field is not too strong (with strong light sources, for example pulsed lasers, higher order terms must be considered, which are the basis of non-linear Raman spectroscopy). The polarizability of a molecule is not a scalar quantity but depends on the relative orientation of the molecule to the electric field vector $\vec{E}(t)$. It is thus expressed as a 3×3 matrix (second-rank tensor) where the element $\alpha_{\rho\sigma}$ determines the contribution from E_σ to μ_ρ ($\rho, \sigma = x, y, z$ in a Cartesian coordinate system).

A molecular vibration displaces nuclei and electron density periodically along the normal coordinate (q), which modulates the polarizability and thus the oscillating dipole moment. Approximating the variation of the polarizability with the kth normal vibration by a series expansion up to the linear term:

$$\underline{\alpha}(q_k) \approx \underline{\alpha}_{\mathrm{eq}} + \left(\frac{\partial \underline{\alpha}}{\partial q_k}\right)_{\mathrm{eq}} q_k(t),$$

and assuming harmonic oscillation of the molecule and electric field:

$$q_k(t) = q_{k_0} \cos(2\pi \nu_k t),$$
$$\vec{E}(t) = \vec{E}_0 \cos(2\pi \nu t),$$

one can derive the following expression [1] from Eq. 2.2:

$$\begin{aligned}\vec{\mu}_k =& \overbrace{\underline{\alpha}_{\mathrm{eq}} \vec{E}_0 \cos(2\pi \nu t)}^{\text{Rayleigh scattering}} \\ &+ \left(\frac{\partial \underline{\alpha}}{\partial q_k}\right)_{\mathrm{eq}} q_{k_0} \vec{E}_0 \frac{1}{2} \left[\underbrace{\cos(2\pi(\nu - \nu_k)t)}_{\text{Stokes scattering}} + \underbrace{\cos(2\pi(\nu + \nu_k)t)}_{\text{anti-Stokes scattering}} \right].\end{aligned} \tag{2.3}$$

Accordingly, the induced dipole moment has three frequency-components, giving rise to Rayleigh, Stokes, and anti-Stokes scattering. The Raman shift ($\pm\nu_k$) depends on the frequency of the molecular vibration, the intensity of the corresponding Raman signal on the square of the partial derivative of the polarizability with respect to

[3] Beside the electric dipole moment, the electromagnetic radiation induces magnetic and higher electric moments, but their effect is several orders of magnitude smaller [1] and can be neglected.

the normal coordinate $(\partial \underline{\alpha}/\partial q_k)_{\text{eq}} = \alpha'_k$ (evaluated at the equilibrium geometry). The simplifications which are employed here are sometimes termed the "double harmonic approximation", which includes the harmonic oscillation of the molecule along the normal coordinate (mechanical harmonicity) and the linear response of the polarizability (electrical harmonicity).

The molecules examined in this work were isolated in jet expansions, where they are oriented randomly relative to the incident radiation. To calculate intensities for randomly oriented molecules, the (derived) polarizability tensor is isotropically averaged over all directions, indicated by $\langle ... \rangle$ in the following. Then, all diagonal elements assume the same (averaged) value, as well as all off-diagonal elements:

$$\langle \alpha'_{xx} \rangle = \langle \alpha'_{yy} \rangle = \langle \alpha'_{zz} \rangle,$$

$$\langle \alpha'_{xy} \rangle = \langle \alpha'_{yx} \rangle = \langle \alpha'_{xz} \rangle = \langle \alpha'_{zx} \rangle = \langle \alpha'_{yz} \rangle = \langle \alpha'_{zy} \rangle.$$

Related quantities can be predicted by means of quantum chemistry. To calculate scattering cross-sections from such predictions, the geometry of the experiment needs to be considered. In the curry-jet setup, a 90° scattering geometry is realized. The light emitted by the laser is highly polarized perpendicular to the scattering plane which is spanned by the light propagation vector (laser beam) and observation vector. The coordinate system depicted in Fig. 2.1 illustrates the scattering geometry of the curry-jet setup and shows some basic components for orientation. In this coordinate system, with the laser polarized perpendicular to the scattering plane [$\perp$(i)], the Stokes-component of the induced dipole moment is:

$$\vec{\mu}_k^{\text{Stokes}} = \frac{1}{2} \begin{pmatrix} \langle \alpha'_{xy} \rangle \\ \langle \alpha'_{xx} \rangle \\ \langle \alpha'_{xy} \rangle \end{pmatrix}_k E_{y_0} \cos\left(2\pi(\nu - \nu_k)t\right). \tag{2.4}$$

Therefore, Raman-scattered light observed at a 90° angle has two components, one polarized parallel to the scattering plane [$\|$(s)], depending on $\langle \alpha'_{xy} \rangle$ (z-component), the other polarized perpendicular to the scattering plane [$\perp$(s)], depending on $\langle \alpha'_{xx} \rangle$ (y-component). Separately measuring the intensity which stems from these components (by using a polarizing filter, for example) and calculating the depolarization ratio provides additional information when assigning molecular vibrations. Although this capability of Raman spectroscopy was not exploited in this work, considering the polarization of the scattered light is relevant when calculated intensities shall be compared to the experiment, because the sensitivity of the setup is not the same for both polarization components and the overall scattering cross-section must thus be weighted according to this deviation.

When calculating components of the derived polarizability tensor, the quantum chemistry software packages used in this work, Gaussian 09 [13] and Turbomole v6.4 [14], do not provide isotropic averages, but the related tensor invariants a' and γ', which do not depend on the orientation of the molecule:

$$\langle \alpha'^2_{xx} \rangle_k = \frac{45a'^2_k + 4\gamma'^2_k}{45},$$

$$\langle \alpha'^2_{xy} \rangle_k = \frac{\gamma'^2_k}{15}.$$

a'_k is the average change of polarizability and γ'_k the change of anisotropy during the vibrational motion along the kth normal coordinate. Gaussian provides a value for the Raman activity $A = 45a'^2 + 7\gamma'^2$ (and polarization ratios) while Turbomole gives separate values for a' and γ'. With these quantities, the differential scattering cross-section suitable to the experimental conditions of the curry-jet (laser polarization and observation angle, photon counting with a CCD detector) can be calculated from the following equation [11]:

$$\sigma'_k = \left(\frac{\mathrm{d}\sigma_k}{\mathrm{d}\Omega}\right) = g_k \frac{2\pi^2 h}{c\tilde{\nu}_k} \cdot \frac{(\tilde{\nu} - \tilde{\nu}_k)^3 \tilde{\nu}}{1 - \exp\left(-\frac{hc\tilde{\nu}_k}{k_B T}\right)} \cdot \left(\underbrace{\frac{45a'^2_k + 4\gamma'^2_k}{45}}_{\perp(\mathrm{s})} + \underbrace{\frac{\gamma'^2_k}{15}}_{\parallel(\mathrm{s})}\right), \quad (2.5)$$

$$\left[\sigma'_k\right] = \mathrm{m}^2\,\mathrm{sr}^{-1},$$

where σ is the total scattering cross-section, Ω the solid angle, $\tilde{\nu}$ the wavenumber of the excitation laser, $\tilde{\nu}_k$ the wavenumber of the kth vibration, and g_k the degeneracy of this vibration. The remaining quantities have their usual meaning.

The exponential term $\left[1 - \exp\left(-hc\tilde{\nu}_k/k_B T\right)\right]^{-1}$ in Eq. 2.5 takes "hot transitions" from excited vibrational states into account, which overlap with the fundamental transition from the ground state in the harmonic approximation, but scatter light more efficiently.[4] Accordingly, the term becomes large if vibrationally excited states are significantly populated, which is the case for vibrations at low wavenumbers and/or high temperatures. If the temperature is sufficiently low or the wavenumber of the vibration sufficiently large, the exponential term approaches one and can be neglected. However, the most interesting vibrations of alkanes and perfluoroalkanes lie at small wavenumbers where the temperature dependence should be considered, if the vibrational temperature is known.

The final step which is necessary to compare calculated scattering cross-sections to the experiment is to invoke the polarization dependent sensitivity of the setup, or more specifically the polarization dependent efficiency of the grating [15] installed in the

[4] This can be rationalized by larger amplitudes (q_{k_0}) of vibrations in higher excited states, which amplifies the Raman term in Eq. 2.3.

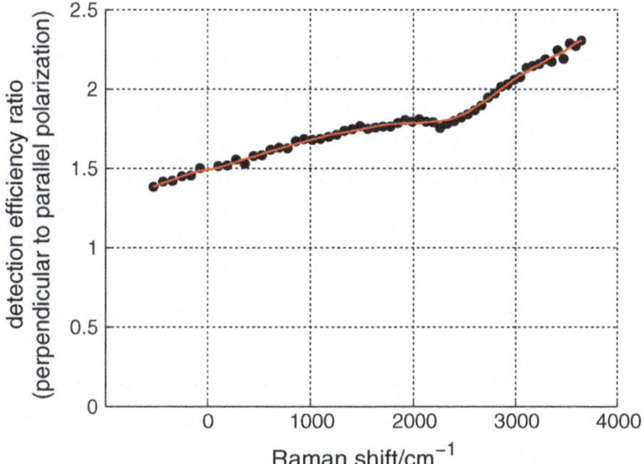

Fig. 2.2 Polarization dependent sensitivity of the curry-jet setup. A tungsten lamp and polarizing filter was used to generate (equally intense) light polarized perpendicular and parallel to the scattering plane. The ratio of the detected light is plotted against the Raman shift (Rayleigh-wavelength = 532 nm) and demonstrates the polarization-dependent detection efficiency. Accordingly, light polarized perpendicular to the scattering plane is detected between ~1.5 and ~2.5 more efficiently than light polarized parallel to the scattering plane, depending on its wavelength

monochromator used to analyze Raman-scattered light.[5] The deviation in sensitivity was characterized by measuring white light from a tungsten lamp with a common polarizing filter[6] placed in front of the monochromator slit. The lamp was placed on the optical axis (negative x-axis in Fig. 2.1) and the intensity was measured by integrating spectra exposed for half a second. The intensity of light polarized parallel and perpendicular to scattering plane defined above was measured in alternation in the wavelength interval $\lambda = 520$–660 nm, which yielded intensity ratios plotted in Fig. 2.2 (plotted against the corresponding Raman shift for the excitation wavelength $\lambda = 532$ nm). The calibration function drawn with a red line provides weighting factors for the terms labeled $\perp$(s) and $\|$(s) in Eq. 2.5, which constitute the two polarization components of the scattering cross-section. Polarization-corrected cross-sections may be compared to experimental intensities and were employed to simulate Raman spectra, which will be discussed later in the text.

---

[5] This step can be avoided when the polarization of Raman-scattered light is scrambled with a polarization scrambler before it enters the monochromator, but this piece of equipment was not available for the measurements presented here and would have reduced the detection sensitivity for polarized transitions.

[6] Measurements were done with two filters: Hama Hoya 450/670 linear polarizing filter or Prinz PO: 012937 polarizing filter. Both yielded the same polarization-dependence within the measurement accuracy (uncertainty of ± 0.1 in the detection efficiency ratio, see Fig. 2.2).

2.2 Free Jet-Expansion

In spectroscopy it is often required to investigate substances at low temperatures, because transitions originating from excited states diminish and the interpretation of spectra becomes much better manageable. Yet, cooling the substance of interest will eventually lead to condensation, and if the molecular surrounding is critical for the property to be examined, investigating a cold solid might not be an option. This is the very case for the alkane-folding study presented in this work: the examined chain molecules possess numerous stable conformational states which compete with the interesting low-energy conformations, such that cooling is crucial to observe the latter, but the preferred conformation depends dramatically on the molecular environment and the "true" conformational preference of a single molecule can only be assessed when it is examined free of interactions with an environment.

One way to escape this dilemma is to "supercool" the gaseous sample, that is, lower the temperature too fast for the rather slow condensation to take place, by preparing a supersonic expansion. A supersonic expansion is formed when a gas is expanded from high pressure (the stagnation pressure, p_0) through a small orifice or nozzle into a vacuum chamber under suitable conditions. This concerns the background pressure (p_b) in the chamber, which must be kept below about half the stagnation pressure, as well as the nozzle dimensions, which must be large enough to allow collisions of the expanded particles in the initial stage of the expansion (larger than the mean free path of the particles at high pressure). If these criteria are met, molecules escaping the reservoir will form a jet characterized by a very narrow velocity distribution, which implies a low translational temperature. The local speed of sound in the expansion decreases proportionally to the square root of the translational temperature, so that the velocity of the particles eventually exceeds the local speed of sound and the expansion may be termed 'supersonic'. In slit jet-expansions, the spectral resolution profits from a narrow velocity distribution along the slit direction, because Doppler broadening is attenuated. This effect can also be achieved or amplified by placing a skimmer in front of the nozzle to collimate the gas flow. The usage of a skimmer is what sets a "molecular beam" apart from a "free jet". With the curry-jet setup, free jet-expansions are generated, as the spectral resolution ($\approx 1\,\text{cm}^{-1}$ [11]) is much poorer than the Doppler broadening.

Heat conduction and viscosity of a supersonic expansion are usually small and the expansion can be treated as being isentropic. Assuming ideal behavior of the expanded gas, one can use isentropic relations to estimate the temperature of the expansion:

$$\frac{p}{p_0} = \left(\frac{T}{T_0}\right)^{\frac{\gamma}{\gamma-1}}.$$

In this work, expansions containing >90 % helium were investigated (heat capacity ratio $\gamma = 5/3$), expanded from commonly $p_0 = 0.5$ bar at $T_0 = 400$ K into a vacuum chamber kept at $p_b \approx 1$ mbar. Plugging these values into the isentropic relation yields a temperature of $T \approx 30$ K.

This is actually an upper bound for the achievable temperature of the supersonic expansion, because the gas is travelling faster than the speed of sound and overexpands to a pressure lower than the background pressure. Tobias Wassermann conducted reference measurements using nitrogen as a probe to deduce the particle density inside a supersonic expansion prepared with the curry-jet [11]. Measurements from a mixture of 1% nitrogen in helium expanded from $p_0 = 1$ bar at $T = 300$ K (nitrogen particle density $\rho_{N_2,0} \approx 2.4 \times 10^{17}\,\text{cm}^{-3}$) yielded a nitrogen particle density of $\rho_{N_2} \approx 6 \times 10^{14}\,\text{cm}^{-3}$ at 1 mm nozzle distance. Assuming that the ratio of N_2 to He remains constant during the expansion, this implies a density drop of $\rho/\rho_0 \approx 2.5 \times 10^{-3}$. Plugging this value into the isentropic relation:

$$\frac{\rho}{\rho_0} = \left(\frac{T}{T_0}\right)^{\frac{1}{\gamma-1}}$$

leads to the translational temperature $T \approx 6$ K (helium has no other relevant degrees of freedom).

Adding a small amount of another substance to a cold helium expansion will not alter the expansion significantly, but the molecules seeded in the cold bath will experience collisions and lose internal energy—they relax into lower energy states. This beneficial relaxation is limited because the density of the expansion quickly decreases down to a point where the collision rate is too small to provide any significant redistribution of energy. In the resulting 'zone of silence', the energy distribution of the molecules on their energy levels freezes while they continue to expand. At some point, the overexpanded gas will collapse under the higher background pressure, which terminates the supersonic expansion. Therefore, the number of collisions is limited (of the order of 100 [8]). Whether an internal degree of freedom reaches the temperature of the bath depends on how efficiently energy is removed by collisions. The rotational temperature determined from the nitrogen measurement mentioned above is $T_{rot} = (19 \pm 3)$ K [11, 16], close to the translational temperature of helium. In general, rotational degrees of freedom can be cooled quite efficiently in supersonic expansions, while vibrational degrees of freedom need more collisions to equilibrate. On top of this, cooling is not equally efficient for all vibrational degrees of freedom, so that they reach different final effective temperatures.

Conformational isomerizations can only take place when collisions with the carrier gas or formation of short-lived complexes [17] provide enough energy to overcome isomerization barriers. Therefore, once the density of the expansion drops below a certain threshold, conformational isomerization ceases and conformational states freeze in a certain distribution [18, 19]. For example, a curry-jet study of gauche- and trans-ethanol yielded a conformational temperature (freezing temperature) of $T_{conf} \approx 120$ K [11, 16], much higher than the rotational temperature of nitrogen expanded under comparable conditions despite a very small effective mass for the isomerization reaction which allows for quantum tunneling contributions.

Conformational temperatures of alkanes determined in this work are largely around 100 K, but not all conformational states freeze at the same temperature—a limitation of the supersonic expansion technique which will be of particular importance in the later analysis.

Rare gases are commonly used to prepare supersonic expansions. They do not possess internal degrees of freedom which would need cooling by themselves and are efficient carrier gases. The cooling performance increases with increasing molar mass [6], but heavier rare gases are more polarizable and tend to form van-der-Waals complexes with the seeded molecules. An excessive condensation of carrier gas ('nanocoating' [16]) on alkanes might interfere with self-solvation, but the temporary formation of short-lived complexes with single particles can help to overcome barriers even at low temperatures [17] and promote the relaxation. Therefore, adding a few percent of a heavier additive to a helium expansion is often enough to significantly enhance the relaxation of still minor molecular compounds [20], and was explored in this thesis with a few small molecular additives. Beside van-der-Waals complexes with the carrier gas, the seeded substance can also form small aggregates or clusters with itself, if the supersonic expansions is prepared in a specific way (for example with a high concentration of the examined substance in the carrier gas, and a high stagnation pressure [7]). This is a powerful way to study intermolecular interactions, but unfavorable for this work, because it might lead to a preference of the all-trans conformation. It was thus tried to minimize cluster formation by keeping the substance concentration low and by working with rather low stagnation pressures.

References

1. D.A. Long, *The Raman Effect: A Unified Treatment of the Theory of Raman Scattering by Molecules* (Wiley, Chichester, 2002)
2. R.L. McCreery, *Raman Spectroscopy for Chemical Analysis* (Wiley, New York, 2000)
3. E. Smith, G. Dent, *Modern Raman Spectroscopy: A Practical Approach* (Wiley, Chichester, 2005)
4. D.H. Levy, Laser spectroscopy of cold gas-phase molecules. Annu. Rev. Phys. Chem. **31**, 197–225 (1980)
5. D.H. Levy, The spectroscopy of very cold gases. Science **214**, 263–269 (1981)
6. J.M. Hayes, Analytical spectroscopy in supersonic expansions. Chem. Rev. **87**, 745–760 (1987)
7. M.D. Morse, in Atomic, Molecular, and Optical Physics: Atoms and Molecules, vol. 29, ed. by F.B. Dunning, R.G. Hulet (Academic Press, Orlando, 1996), Part B, pp. 21–47
8. M. Herman, R. Georges, M. Hepp, D. Hurtmans, High resolution fourier transform spectroscopy of jet-cooled molecules. Int. Rev. Phys. Chem. **19**, 277–325 (2000)
9. C.J. Cramer, *Essentials of Computational Chemistry: Theories and Models*, 2nd edn. (Wiley, Chichester, 2004)
10. F. Jensen, *Introduction to Computational Chemistry*, 2nd edn. (Wiley, Chichester, 2007)
11. T.N. Wassermann, Umgebungseinflüsse auf die C-C- und C-O-Torsionsdynamik in Molekülen und Molekülaggregaten: Schwingungsspektroskopie bei tiefen Temperaturen, Ph.D. Thesis, Georg-August-Universität Göttingen, 2009
12. C.V. Raman, K.S. Krishnan, A new type of secondary radiation. Nature **121**, 501–502 (1928)
13. M.J. Frisch, G.W. Trucks, H.B. Schlegel, G.E. Scuseria, M.A. Robb, J.R. Cheeseman, G. Scalmani, V. Barone, B. Mennucci, G.A. Petersson, H. Nakatsuji, M. Caricato, X. Li, H.P.

Hratchian, A.F. Izmaylov, J. Bloino, G. Zheng, J.L. Sonnenberg, M. Hada, M. Ehara, K. Toyota, R. Fukuda, J. Hasegawa, M. Ishida, T. Nakajima, Y. Honda, O. Kitao, H. Nakai, T. Vreven, J.A. Montgomery, J.E. Peralta, F. Ogliaro, M. Bearpark, J.J. Heyd, E. Brothers, K.N. Kudin, V.N. Staroverov, R. Kobayashi, J. Normand, K. Raghavachari, A. Rendell, J.C. Burant, S.S. Iyengar, J. Tomasi, M. Cossi, N. Rega, J.M. Millam, M. Klene, J.E. Knox, J.B. Cross, V. Bakken, C. Adamo, J. Jaramillo, R. Gomperts, R.E. Stratmann, O. Yazyev, A.J. Austin, R. Cammi, C. Pomelli, J.W. Ochterski, R.L. Martin, K. Morokuma, V.G. Zakrzewski, G.A. Voth, P. Salvador, J.J. Dannenberg, S. Dapprich, A.D. Daniels, O. Farkas, J.B. Foresman, J.V. Ortiz, J. Cioslowski, D.J. Fox, Gaussian 09, Revision A.02 (Gaussian Inc, Wallingford CT, 2009)

14. TURBOMOLE v6.4 2012, A development of University of Karlsruhe and Forschungszentrum Karlsruhe GmbH, 1989–2007, TURBOMOLE GmbH, since 2007; available from http://www.turbomole.com
15. A. Hessel, A.A. Oliner, A new theory of wood's anomalies on optical gratings. Appl. Opt. **4**, 1275–1297 (1965)
16. T.N. Wassermann, M.A. Suhm, Ethanol monomers and dimers revisited: a Raman study of conformational preferences and argon nanocoating effects. J. Phys. Chem. A **114**, 8223–8233 (2010)
17. U. Erlekam, M. Frankowski, G. von Helden, G. Meijer, Cold collisions catalyse conformational conversion. Phys. Chem. Chem. Phys. **9**, 3786–3789 (2007)
18. P. Felder, H.H. Günthard, Conformational interconversions in supersonic jets: matrix IR spectroscopy and model calculations. Chem. Phys. **71**, 9–25 (1982)
19. R.S. Ruoff, T.D. Klots, T. Emilsson, H.S. Gutowsky, Relaxation of conformers and isomers in seeded supersonic jets of inert gases. J. Chem. Phys. **93**, 3142–3150 (1990)
20. W.Y. Sohn, J.S. Kang, S.Y. Lee, H. Kang, Fluorescence excitation spectrum and solvent-assisted conformational isomerization (SACI) of jet-cooled acetaminophen. Chem. Phys. Lett. **581**, 36–41 (2013)

Chapter 3
Experimental

This chapter will outline the experimental setup as well as the development and layout of a heatable nozzle, required to prepare longer alkanes in the gas phase (Sect. 3.1). Details on the basic components of the curry-jet are summarized in Table 3.1. Details on the heatable nozzle can be found in Table 3.2. Furthermore, it will be shown how the variable measurement conditions influence spectra from n-alkanes (Sect. 3.2). The spectra will not yet be analyzed in detail at this point, but some assignments will be anticipated to discuss the observed effects.

3.1 Curry-Jet Setup

The curry-jet is a versatile Raman spectrometer which allows to probe cold and nearly isolated molecules due to the free-jet preparation technique. The first version of this setup was built by Philipp Zielke [1] and later improved by Tobias Wassermann [2] and Zhifeng Xue [3]. In the beginning of this work, the characterization of small volatile substances and aggregates with the curry-jet and its predecessor was already well established (see for example Refs. [4–9]) and the access to larger, low-volatile molecules was desired [10]. To this end, efforts were made to equip the curry-jet with a heated nozzle [11–15], where heating of the substance would increase its vapor pressure and provide a sufficient concentration in the carrier gas (see Eq. 2.1). The design of a well-performing heatable nozzle was the central technical task of this work. A sketch of the curry-jet is shown in Fig. 3.1. The setup was used as described in detail by Tobias Wassermann [2] with exception of the heatable nozzle and smaller improvements regarding stray light. The general setup organization will thus be outlined here more briefly and the focus will be kept on the new equipment.

In the previous "room temperature version", the curry-jet is fed with the substance of interest by letting an inert and Raman-inactive carrier gas pass through a glass saturator which is operated between −30 °C and room temperature (saturator-temperature $\equiv \vartheta_S$). The carrier gas picks up some substance held by the saturator and the mixture is temporarily stored in a reservoir at room temperature from where it can be fed to the nozzle and be expanded into the vacuum chamber without significant variation in stagnation pressure during the long gas pulse. Adjustment of the gas flow

N.O.B. Lüttschwager, *Raman Spectroscopy of Conformational Rearrangements at Low Temperatures*, Springer Theses, DOI 10.1007/978-3-319-08566-1_3

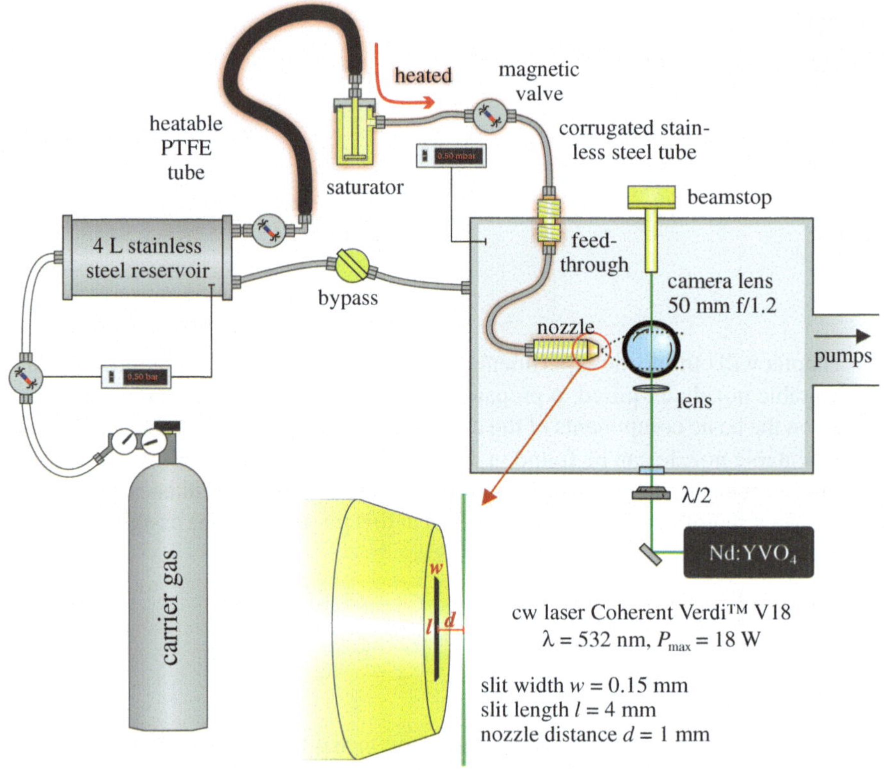

Fig. 3.1 Sketch of the curry-jet setup (heatable version). Heating is provided by resistance heating cables or wires, wrapped around the components. Reference II—Reproduced by permission of The Royal Society of Chemistry

is achieved by pressure gauges which control magnetic valves. The temperature of the glass saturator may not exceed room temperature, or else the substance mixed with the carrier gas will reach saturation and condense in colder supply tubes, valves, the reservoir, or nozzle. In the heatable version, which is depicted in the figure, a brass flow-through saturator wrapped with heating cable is used to mix the substance of interest with the carrier gas. All equipment from the saturator up to the nozzle can be operated at a temperature up to $\approx$150 °C,[1] which allows to heat the examined substance up to $\approx$130 °C. A small temperature difference is kept because some spots of the heated supply assembly are colder than the set-temperature, especially couplings connecting tubes and valves. As for the saturator, heating is provided by resistance heating cables or wires, wrapped around corrugated stainless steel tubes, the magnetic valve controlling the gas flow to the nozzle, and the nozzle. The temperature is measured with several thermocouples at selected positions and heating is regulated

[1] Higher temperatures would lead to degeneration of sealing Viton® O-rings.

Table 3.1 Main components of the curry-jet setup

Excitation and Detection
Laser: Verdi™ V18 (Coherent, Inc.) frequency doubled Nd:YVO_4 cw laser, single mode emission at 532 nm, 18 W maximum output power, linewidth $< 0.0002\,\text{cm}^{-1}$
Camera lens: Nikkor 50 mm $f/1.2$ (Nikon GmbH) 50 mm focal length, maximum aperture ratio = 1.2
Rayleigh-filter: REFUS532-25 (LOT-QuantumDesign GmbH) long-pass edge filter, ultra-steep, low-ripple, transmission $> 90\,\%$ for $\lambda > 533.7\,\text{nm}$ ($\tilde{\nu} > 60\,\text{cm}^{-1}$), optical density = 6 at $\lambda = 532\,\text{nm}$, diameter = 25 mm
Monochromator: Model 2501 (McPherson, Inc.) Czerny-Turner design, 1 m focal length, aperture ratio = 8.7, ruled grating with 1200 grooves/mm. Overall spectrometer resolution $\approx 1\,\text{cm}^{-1}$ [2]
Detector: Spec-10: 400B/LN (Princeton Instruments) back-illuminated CCD detector with 1340 × 400 pixel ($26.8 \times 8.0\,\text{mm}^2$), operation temperature: −120 °C (liquid nitrogen cooled), quantum yield $> 90\,\%$ for $\lambda = 500$–700nm, controlled by a Windows PC running WinSpec[a], operated in full vertical binning mode
Preparation and Vacuum
Reservoir: custom built $4.7\,\text{dm}^3$, stainless steel or PTFE coated
Vacuum chamber: custom built $40 \times 60 \times 60\,\text{cm}^3$, anodized aluminum
Roots pumps: Okta series, WKP 500 AM and WKP 250 AM (Pfeiffer Vacuum GmbH) $500\,\text{m}^3\,\text{h}^{-1}$ and $250\,\text{m}^3\,\text{h}^{-1}$ pump capacity
Rotary vane pump: UNO 101 S (Dr.-Ing. K. Busch GmbH) $100\,\text{m}^3\,\text{h}^{-1}$ pump capacity

[a]http://www.princetoninstruments.com/products/software/ (16 September 2013)

by PID controllers. A ≈3 m long polytetrafluoroethylene (PTFE or Teflon®) coated hose is used to pre-heat the carrier gas before it enters the saturator. The reservoir does not store the carrier gas/substance mixture in the heated version, because a uniform temperature control would be very demanding for this rather large object, but carrier gas additives, such as argon or tetrafluoromethane (see below) can be mixed to the carrier gas over additional couplings on the reservoir, controlled by magnetic valves and an oscilloscope to monitor opening times.

All experiments presented here made use of a $(4 \times 0.15)\,\text{mm}^2$ slit nozzle pumped by two roots pumps and one rotary vane backing-pump. The vacuum chamber hosts a plano-convex lens focusing the laser beam (focused beam diameter ≈30 μm) and a camera lens to collect scattered light in a 90° observation-angle (in the figure, the camera lens points perpendicular to the paper plane). The relative position of the camera lens and nozzle in *xyz*-directions (Fig. 2.1) can be set with precise motorized actuators. In addition, the absolute distance d_n from the nozzle to the laser beam, which crosses the expansion at an angle of 90°, can be controlled with a precision

Table 3.2 Components of the heated nozzle

Carrier gas heating hose: H1201-40-8 (HORST GmbH) length ≈ 3 m, PTFE coated, heatable up to 200 °C, 220 V, 560 W, type J thermocouple (Fe-CuNi), HT M2 temperature controller (0–200 °C) *Heating cables*: HSS – 450 °C (HORST GmbH) with protective braid, length = 0.6–2.0 m (75–250 W, 230 V), diameter = 3.5–4.5 mm, minimal bending radius = 6 mm, maximal temperature 450 °C
Heating controller (I): 3216 temperature controller (Eurotherm) PID controller, accuracy = ±0.1 °C (temperature-control of nozzle and saturator)
Heating controller (II): E6C 1/16 DIN temperature controller (RS Components GmbH) PID controller, accuracy = ±1 °C (temperature-control of supply tubes, feed-through, and magnetic valve)
Heating wires: SEA series (Thermocoax) stainless steel sheath, diameter = 1.5 mm, minimal bending radius = 3 mm, 100 W (length = 1 m, operated at 24 V), maximal temperature 600 °C
Magnetic valve: Buschjost series, 85572100.9402.02400 (Rauh Hydraulik GmbH) 2/2-way magnetic valve, normally closed, brass, PTFE seals, −20–200 °C, 24 V, 29 W, G 3/8″ connection
Temperature sensors: custom built type K thermocouples (NiCr-Ni)

of <0.03 mm. Scattered light collimated by the camera lens passes through a window to a collecting lens outside the vacuum chamber which focuses it onto the slit of a Czerny-Turner monochromator equipped with a liquid nitrogen cooled, sensitive CCD detector. A Raman edge filter positioned in front of the monochromator slit attenuates light at the excitation wavelength by six orders of magnitude. In order to suppress unwanted stray light further, adjustable anodized aluminum tubes placed inside the vacuum chamber encompass the laser beam and collimated beam of collected light. Additionally, the lens focusing the laser beam is covered with an iris diaphragm. Especially the latter helps to suppress stray light reflected from the nozzle, which notably increases the signal-to-noise ratio in spectra close to the Rayleigh-line. Since this is the region were the most interesting alkane vibrations are found, these simple improvements were actually important for the success of this work.

The nozzle and detection optics are aligned prior to each measurement by expanding ambient air and maximizing Raman signals from nitrogen. Spectra are then recorded from continuous carrier gas/substance expansions. Commonly, the CCD detector is exposed between one to ten minutes. Longer exposure times are possible but rarely chosen, because the spectra start to suffer from cosmic rays which cause intense sharp signals ("spikes") covering the spectrum in a random fashion. To identify and remove spikes, every spectrum is recorded at least four times. A small script written in Matlab [16] compares single "scans" and removes spikes based on the following algorithm [1]: For each data point of the spectrum, the average intensity over all scans is calculated and compared to the intensity measured in each scan. Data points will be classified as spikes if they are larger than the mean value by an adjustable threshold. When a spike is identified, it is overwritten by the average value from the previous and following scans which are not identified as spikes. This procedure is repeated iteratively with decreasing thresholds and successfully removes spikes in the vast majority of spectra.

Emission lines from a neon lamp serve as wavelength reference to calibrate recorded Raman spectra. The neon lamp is placed on the optical axis defined by the focus of the excitation laser beam, detection optics, and monochromator slit, such that the light crosses the vacuum chamber (negative x-axis in Fig. 2.1). In addition to the wavelength-calibration, the lamp is used to determine the reference point for the nozzle distance ($d_n = 0$) [2]: After filling the vacuum chamber with air and optimizing the detection optics to a Raman measurement, neon emission lines are monitored with the laser switched off. The nozzle is driven forward until the intensity drops to half of its initial value, which happens within a displacement interval of <0.03 mm and determines the $d_n = 0$ position accurately [2]. This procedure is safer than driving the nozzle into the laser beam, which could cause stray light to damage the CCD detector even at low laser output-power.

3.1.1 Heatable Nozzle and Saturator

Like many other parts of the curry-jet, the heatable nozzle was custom built by the mechanics workshop of the Institute of Physical Chemistry in Göttingen. An assembly drawing is depicted in Fig. 3.2. The main parts are a threaded inner stainless steel tube which is inserted into two more massive brass cylinders wrapped with heating wires (later heating cables), and an interchangeable nozzle-attachment which is positioned on two pins on top of the inner tube and hold in place by a retainer ring. The nozzle is sealed with a Viton® O-ring. Prior to this work, experiments were made with a pot connected perpendicularly to the inner tube of the nozzle over a $\approx$3 cm long tube with an inner diameter of 4 mm (Fig. 3.2a). It was anticipated that a sufficient amount of the substance evaporating in the pot would diffuse into the nozzle and get dragged away by the carrier gas. This approach has the advantage of simple implementation and low demands regarding temperature control, because the only part being heated was the nozzle itself, the pot was heated passively by heat transfer. Disadvantages of this approach were that the pot could not be sealed. The substance was thus continuously exposed to vacuum while the apparatus was running and was lost while no measurement was made. Also, it was not possible to set the substance temperature independently from the nozzle temperature. Most importantly, this design failed to provide sufficiently concentrated carrier gas/substance mixtures, which was attributed to insufficient diffusion of substance through the rather undersized tube connecting nozzle and pot.

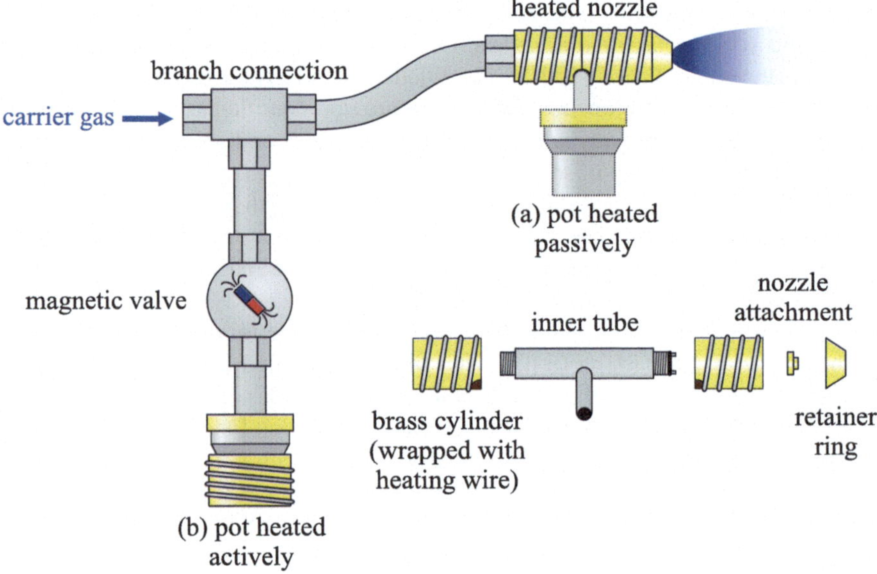

Fig. 3.2 Early designs of the heatable nozzle

First experiments from this work were made with the pot being connected over tubes with a larger inner diameter of 8 mm. Based on the Hagen-Poiseuille equation,[2] the doubled tube diameter was expected to allow an increased transport of substance of at least one order of magnitude. Some of the drawbacks of the earlier design like substance loss and inflexible substance/nozzle temperatures were tackled by connecting the pot not directly to the nozzle but to flexible supply tubes over a branch connection and magnetic valve (Fig. 3.2b). All these components including the pot were heated with heating cables and placed inside the vacuum chamber to avoid the need of a heated feed-through.

As its predecessor, this design failed to provide satisfactory substance concentrations. Test measurements with octane revealed that a sufficient concentration could only be reached at stagnation pressures between 50–100 mbar—one order of magnitude smaller than stagnation pressures chosen in previous routine measurements. Consistently, the octane spectra were indicative of a high expansion temperature, which could not be explained with the nozzle being heated: the glass saturator was found to perform better in terms of concentration *and* expansion temperature at the same nozzle temperature, but higher stagnation pressure.

As a consequence of these findings, designs where the substance needs to diffuse into the carrier gas were dropped. The best accumulation of substance in the carrier

[2] The Hagen-Poiseuille equation states that the volume transport of a laminar flow through a cylindrical pipe is proportional to the 4th power of the pipe's radius.

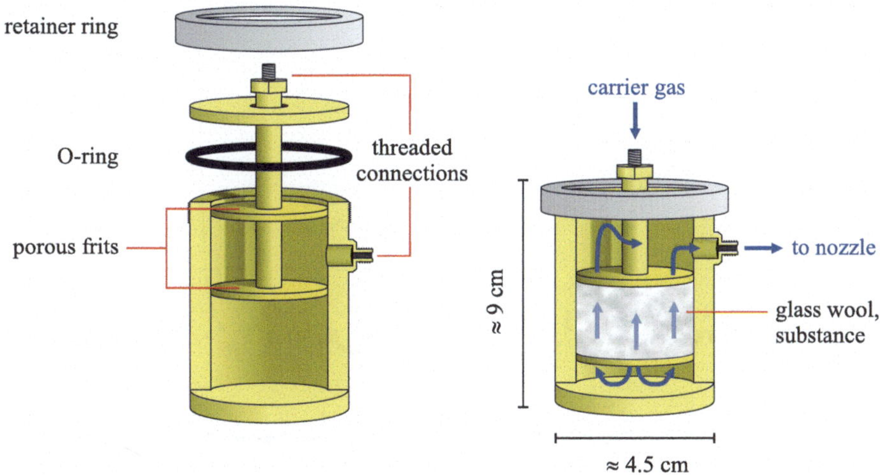

Fig. 3.3 Exploded assembly drawing of the brass saturator. Heating is achieved with a heating cable wrapped around the saturator (not shown). Reference II—Reproduced by permission of The Royal Society of Chemistry

gas was to be expected with a flow-through system like the glass saturator, which is why this design was transfered to the heatable brass saturator depicted in Fig. 3.3. Additionally, a heatable feed-through was constructed, allowing to place the saturator outside the vacuum chamber, much more conveniently accessible to refill substance between measurements. This gave the finally working "heated version" of the curry-jet, which was introduced above and is depicted in Fig. 3.1. With this setup, it was possible to measure low-volatile alkanes up to heneicosane (21 carbon atoms). A comparison of hexadecane spectra in Fig. 3.4 demonstrates the potential of substance preparation above room temperature.

During research with the working version of the heatable nozzle, some further shortcomings became apparent, which shall be outlined in the following. A general problem with the heated nozzle present in all versions is that the mountings which hold the nozzle and optics are subject to passive heating which causes the alignment to drift. Measurements with the heated nozzle thus involve a longer warm-up period where the setup is allowed to thermally equilibrate, and realignment is necessary more often than in case the setup is run at room temperature. Another problem concerns the preparation of the substance. As indicated in Fig. 3.3, the substance is filled into the brass saturator together with some inert glass wool. This has the advantage of providing a larger surface to the substance and thus promoting its accumulation in the carrier gas, but the glass wool was initially added because the carrier gas presses the substance out of the saturator otherwise, probably because the substance forms bubbles which can easily ascend in the small saturator. Glass wool was found to be a better substrate than molecular sieve, which rather inhibits the accumulation of substance in the carrier gas, but it contains very small particles which can be carried

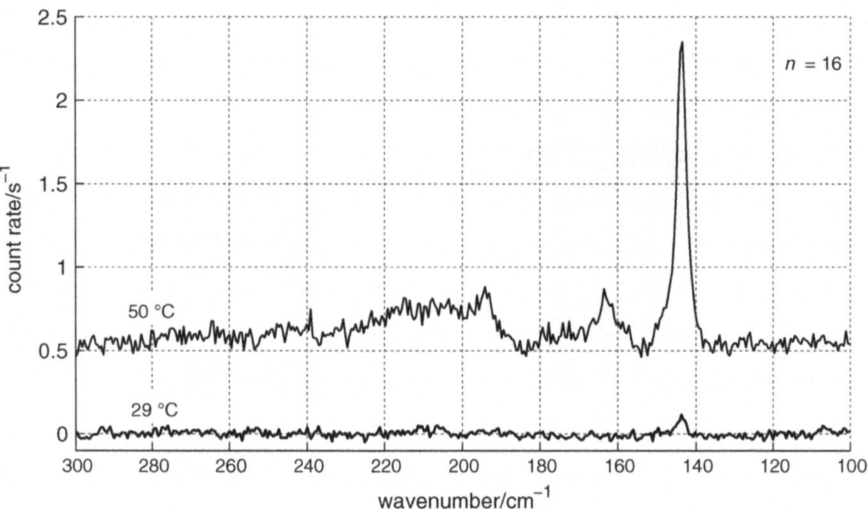

Fig. 3.4 Raman jet spectra of hexadecane, prepared at room temperature with the glass saturator (*lower curve*), and at 50 °C with the heatable brass saturator (*upper curve*). The room temperature spectrum was measured with a total exposure time of 10 h, at the edge of detectability, while the measurement at 50 °C took 30 min

off by the carrier gas and block the nozzle. This is not always obvious and may alter the expansion conditions.

Commonly, the concentration decreases in the course of a measurement, in particular in the initial phase. The effect can be dramatic when the flow of the carrier gas through the saturator is temporarily interrupted, for example to record a background spectrum, which was thus done prior to each measurement. To check for impurities, background spectra were measured from pure helium expansions in any case, because condensation of alkanes in supply tubes could not fully be avoided. Especially measurements of longer alkanes with chain lengths $n > 17$ often made intense cleaning of supply tubes and optics necessary. Finally, heating cables placed inside the vacuum chamber could not be operated at 230 V, because residual helium at low-pressures caused a glow discharge surrounding the cables which interfered with the detection of Raman signals. It was thus necessary to keep the voltage below ≈180 V to avoid the glow discharge of the carrier gas, which hampered the heating performance a little.

Some parts of the heatable nozzle are currently being revised to reduce the discussed problems. The heatable brass saturator will be replaced by a longer version which will (hopefully) render the use of a substrate unnecessary. Furthermore, a temperature-resistant four-way valve with two L-shaped ports will be installed to easily switch between the glass and brass saturator.

It is worth mentioning that, beside giving the opportunity to measure low-volatile substances, a heatable nozzle is also a valuable asset when investigating small aggregates. Higher nozzle temperatures inhibit cluster formation and thus support the assignment from cluster bands, which deplete when the nozzle temperature is gradually increased [17]. This is a convenient side-effect, because cluster formation is unfavorable for the alkane-folding study. Furthermore, variation of the stagnation temperature can be useful to control the starting distribution of conformations, which is advantageous when conformational conversion freezes at relatively high temperature.

3.2 Measurement Conditions and Influence on Jet Spectra

Among the different alkanes, measurement conditions were kept as comparable as possible, in order not to superimpose artificial spectral evolutions over true chain length dependent trends. The mole fraction of the alkanes in the carrier gas (x) is one aspect which needs attention, because it can affect the amount of cluster formation in jet expansions. It is controlled by the saturator temperature (ϑ_{S}) which determines the vapor pressure (p_{V}) of the filled substance, as well as the applied pressure of the carrier gas flushed through the saturator (p_0), and estimated for an ideal gas mixture according to:

$$x = \frac{p_{\mathrm{V}}(\vartheta_{\mathrm{S}})}{p_0}, \tag{3.1}$$

which assumes that the vapor pressure is not altered by the presence of the carrier gas and equilibrium is always reached. The substance probably evaporates slower than it is dragged away by the carrier gas, so that the true mole fraction is less than given by Eq. 3.1.

Measurements in the chain length interval $n = 13$–21 were carried out using the heatable brass saturator and nozzle described above. Mixtures of shorter alkanes were prepared with the glass saturator or prior to the measurement in a vacuum line. For $n = 13$–21, the saturator temperature was set to provide a vapor pressure of $p_{\mathrm{V}} \approx 0.1$ mbar. Figure 3.5 shows the temperatures necessary to reach these vapor pressures for $n = 13$–20 [18] and emphasizes the need for a preparation above room temperature. With a stagnation pressure chosen between $p_0 = 0.5$–1.0 bar, mole fractions were in the per mille region—a compromise between sufficient signal-to-noise ratio and minimized cluster formation, which are both favored at higher concentrations. Occasionally, the temperature was re-adjusted to balance variations in the mole fraction.

Aside from a uniform mole fraction provided for all chain lengths, parameters which govern the expansion properties and thereby the conformational distribution are particularly important. These parameters are the carrier gas composition, stagnation pressure, and nozzle temperature. In addition, the adjustable distance from the nozzle to the excitation laser can be used to probe different regions of the expansion having different effective temperatures. Increasing the nozzle distance, one allows for more efficient cooling, because the seeded molecules may experience more collisions

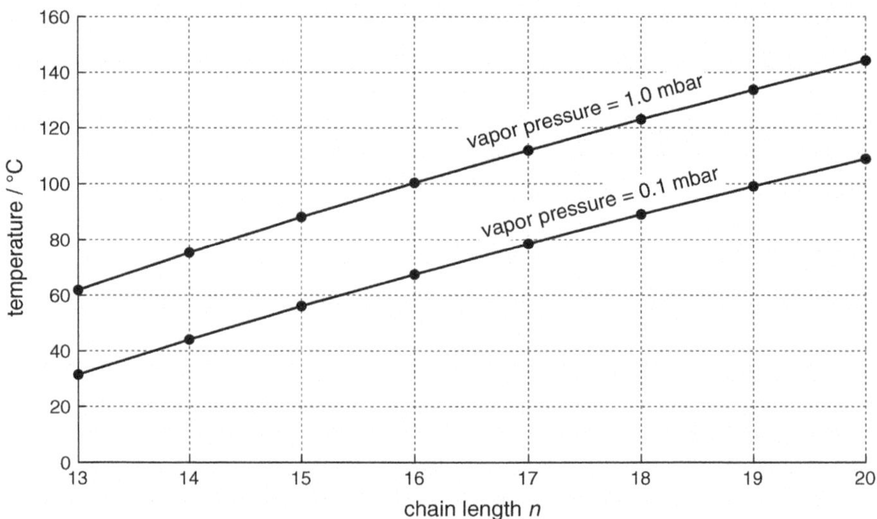

Fig. 3.5 Required temperatures to achieve a vapor pressure of $p_V = 1.0$ mbar and $p_V = 0.1$ mbar, respectively, for the n-alkanes tridecane to eicosane [18]

with the carrier gas before they are probed. This effect can aid the assignment of low energy species [9] or hot bands [7]. The nozzle geometry does also influence the expansion properties, but it was not varied.

3.2.1 Nozzle Distance

In principle, adjustment of the nozzle distance is suited to distinguish alkane conformers [2], but modifying the nozzle distance to both, a relatively short or long setting, has disadvantages compared to a distance of 1 mm, which was used in this work for the most part. The main drawback when increasing the nozzle distance stems from the decreased density of the probed expansion. The consequential weaker Raman scattering needs to be compensated by longer exposure times. This is prevented at longer chain lengths, because low-volatile alkanes with $n > 18$ start to deposit on the detection optics, leading to signal loss and increased stray light. On top of this, unwanted cluster formation is found to occur at longer nozzle distances [2]. In Fig. 3.6, a mild increase of the nozzle distance is demonstrated for hexadecane. The corresponding measurements in helium expansions were carried out consecutively with equal experimental parameters except the nozzle distance. The remaining deviations after background subtraction do not cover the fact that the abundance of the all-trans conformer (reflected by the intensity of the accordion vibration at $\approx 140\,\text{cm}^{-1}$) is affected very little when increasing the nozzle distance by 50 % (as opposed to changing the

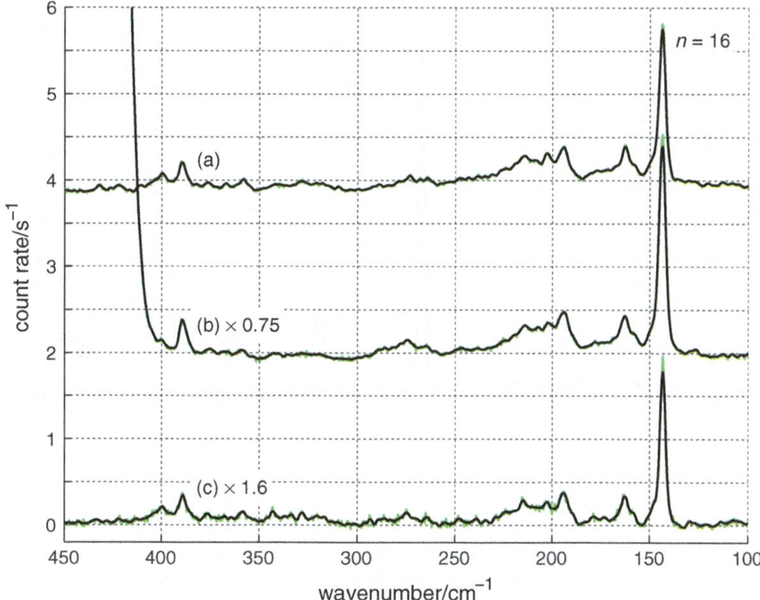

Fig. 3.6 Raman jet spectra of hexadecane in the low-frequency region: **a** He expansion, 1 mm nozzle distance, **b** He + 4 % CF_4 expansion, 1 mm nozzle distance, and **c** He expansion, 1.5 mm nozzle distance. The heavier carrier gas additive (**b**) is seen to increase the relaxation to all-trans, but in case of helium expansions, increasing the nozzle distance from 1 mm (**a**) to 1.5 mm (**c**) has little effect. The intense signal at $\tilde{\nu} > 400\,\text{cm}^{-1}$ in (**b**) is a CF_4 deformation vibration. Original data is plotted in *green*, Savitzky-Golay filtered spectra (9 pt.) in *black*. Spectra are scaled on the bands at ≈160 and ≈190 cm^{-1}. *Measurement conditions* **a** $\vartheta_s = 65\,°\text{C}$, $p_0 = 0.5$ bar, $p_b = 0.9$ mbar, $d_n = 1$ mm, exposure 12 × 300 s, **b** $\vartheta_s = 65\,°\text{C}$, $p_0 = 0.9$ bar, $p_b = 1.3$ mbar, $d_n = 1$ mm, exposure 6 × 300 s, and **c** $d_n = 1.5$ mm, exposure 6 × 450 s, otherwise same as (**a**)

carrier gas composition, see below). Decreasing the nozzle distance can enhance contributions from energetically less favorable species and in that way support the assignment, but suffers from increased stray light, reflected from the nozzle. Also, a shorter nozzle distance will not allow to work out energetically favored conformers, disguised by entropically favored gauche species.

3.2.2 Nozzle Temperature

Another factor which impacts the effective temperature of the expansion is the nozzle temperature. It was commonly set in the range 120–140 °C (mostly 130 °C). Deviations from this setting will be emphasized, when applicable. This rather high temperature was necessary for the measurements of the longest alkanes considered ($n = 20, 21$) and retained for shorter alkanes, not to superimpose an additional

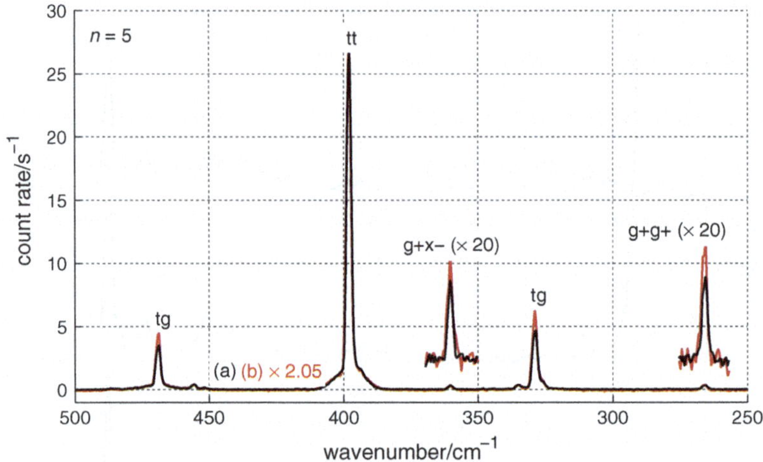

Fig. 3.7 Raman jet spectra of low-frequency pentane vibrations at different nozzle temperatures. The labels mark the all-trans conformer (tt) and gauche conformers (tg, g^+g^+, g^+x^-, see Sect. 4.1). *Measurement conditions* **a** 0.1% in He (prepared in a vacuum line), $p_0 = 0.5$ bar, $p_b = 0.9$ mbar, $d_n = 1$ mm, nozzle at room temperature, exposure 6×300 s, **b** nozzle at 130 °C, otherwise same as (**a**)

effect on the spectral evolution with chain length. The effect of nozzle heating was tested with pentane expanded in helium (Fig. 3.7), where bands from the all-trans conformer and gauche conformers are straightforwardly identified (Ref. [19] and quantum chemical prediction, see below) and do not overlap. Changing the nozzle temperature from room temperature to 130 °C leads to a moderate increase of the gauche conformer abundance as expected, which is a small drawback, but unavoidable. The slightly weaker gain of the tg band at $\approx$470 cm^{-1} with heating relative to the tg band at $\approx$330 cm^{-1} is consistent with a smaller enthalpy difference obtained in cryosolutions [20]. It may be due to some spectral overlap. A comparison to quantum chemical calculations (see Sect. 4.3) suggests that the conformational temperature of pentane expanded in helium is $T_{\text{conf}} \approx 170$K in case the nozzle is kept at room temperature, and $T_{\text{conf}} \approx 200$ K when the nozzle is heated to 403 K (130 °C). A $\approx$30 % transfer of the nozzle temperature increase to the conformational temperature has been found before in the case of hexafluoroisopropanol [21].

3.2.3 Carrier Gas and Stagnation Pressure

To increase the jet cooling efficiency, a way better suited to the demands of this work than varying the nozzle distance is to vary the composition of the carrier gas and the stagnation pressure. The majority of experiments presented here included alkanes expanded in helium, helium/argon ($\approx$6 % Ar), and helium/tetrafluoromethane

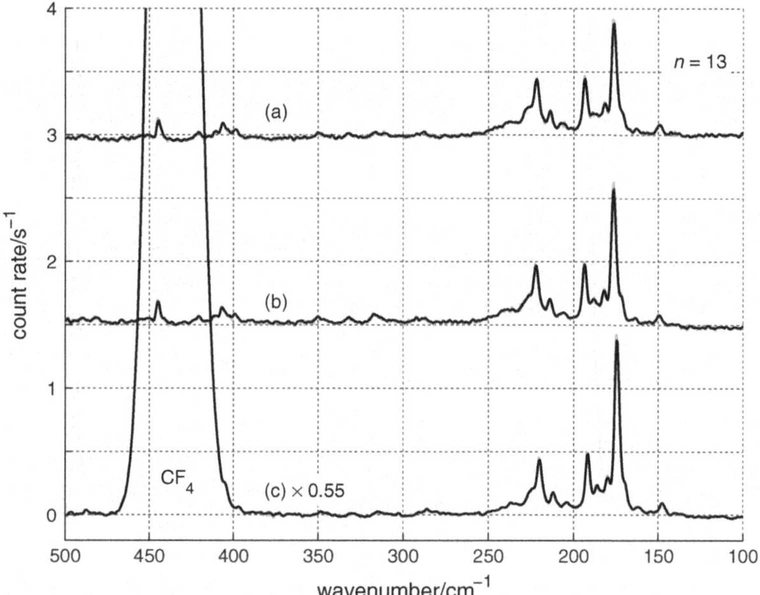

Fig. 3.8 Spectra of the low-frequency region of tridecane in expansions with different carrier gas compositions but otherwise similar experimental parameters. **a** He expansion, **b** He + 6 % Ar expansion, and **c** He + 4 % CF_4 expansion. Original data is plotted in *green*, Savitzky-Golay filtered spectra (9 pt.) in *black*. *Measurement conditions* **a** $\vartheta_s = 37\,°C$, $p_0 = 0.5$ bar, $p_b = 0.9$ mbar, $d_n = 1$ mm, exposure 24 × 300 s, **b** $\vartheta_s = 39\,°C$, $p_0 = 0.8$ bar, $p_b = 1.4$ mbar, $d_n = 1$ mm, exposure 24 × 300 s, **c** $\vartheta_s = 37\,°C$, $p_0 = 0.9$ bar, $p_b = 1.3$ mbar, $d_n = 1$ mm, exposure 24 × 300 s

(≈4 % CF_4). Fewer measurements were done with mixtures of helium/hexafluoroethane (C_2F_6), helium/sulfur hexafluoride (SF_6), and helium/neon. In each case, helium expansions doped with such additives provided increased collisional cooling compared to pure helium expansions. Especially tetrafluoromethane was found to improve the cooling efficiently and was used routinely in subsequent measurements to lower the temperature of co-expanded alkanes.

In Fig. 3.8, the effect of the carrier gas additives argon and tetrafluoromethane is demonstrated on the basis of tridecane spectra. The helium expansion was prepared from a stagnation pressure of ≈0.5 bar, helium/argon and helium/tetrafluoromethane expansions from a stagnation pressure of ≈0.8–0.9 bar to further increase collisional cooling. These pressure settings were used routinely on subsequent measurements. The spectra are scaled on the bands from gauche conformers (193,222 cm^{-1}), such that the accordion vibration band from the all-trans conformer (176 cm^{-1}) increases at lower conformational temperatures. Adding argon to the expansion and increasing the stagnation pressure has a rather small effect on the conformational temperature, but replacing argon by the heavier tetrafluoromethane (a little more than two times the mass of argon, 88 and 40 g mol^{-1}, respectively) enhances the relaxation to all-trans substantially—the accordion band peak intensity increases by about 60 % relative

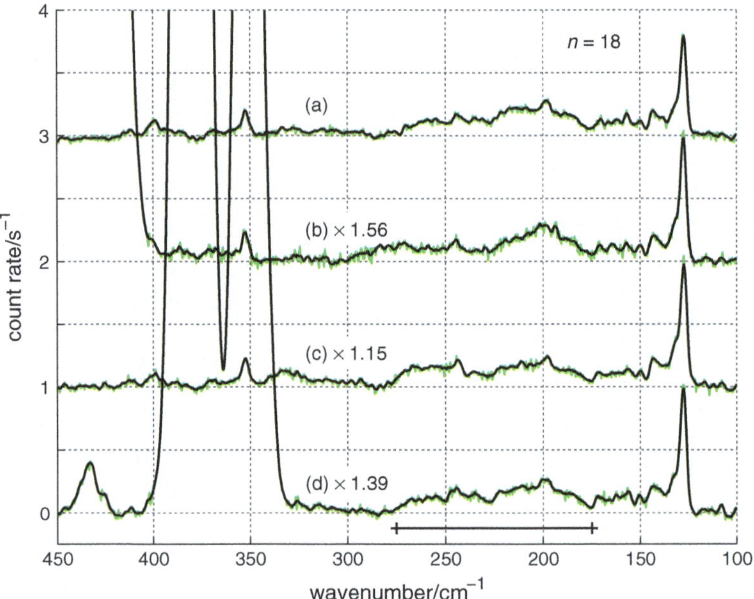

Fig. 3.9 Spectra of the low-frequency region of octadecane in expansions with different carrier gas compositions but otherwise similar experimental parameters. **a** He expansion, **b** He + 4 % CF_4 expansion, **c** He + SF_6 expansion, and **d** He + C_2F_6 expansion. Spectra are scaled with respect to the broad band at $\approx 175 - 275\,\text{cm}^{-1}$ (the integration interval is indicated by a *black line*). Cut-off bands stem from CF_4 (**b**) and C_2F_6 (**d**). Original data is plotted in *green*, Savitzky-Golay filtered spectra (9 pt.) in *black*. *Measurement conditions* **a** $\vartheta_s = 97\,°\text{C}$, $p_0 = 0.5\,\text{bar}$, $d_n = 1\,\text{mm}$, exposure $6 \times 300\,\text{s}$, **b** $\vartheta_s = 97\,°\text{C}$, $p_0 = 0.9\,\text{bar}$, $d_n = 1\,\text{mm}$, exposure $4 \times 400\,\text{s}$, **c** exposure $6 \times 300\,\text{s}$, otherwise same as (**b**), and **d** exposure $8 \times 300\,\text{s}$, otherwise same as (**b**)

to the measurement in pure helium. The same is true for hexadecane (Fig. 3.6) and the other examined alkanes. At higher chain length, the relaxation is somewhat less pronounced, but still reaches 20–40 %. The use of tetrafluoromethane as carrier gas additive can be recommended if its (Raman) spectrum does not cover signals of interest. Characteristic alkane vibrations are not covered by CF_4 bands, which served as additional wavenumber reference.

Experiments with sulfur hexafluoride (SF_6) and hexafluoroethane (C_2F_6) as carrier gas additives were restricted to octadecane. Spectra of the low-frequency region are shown in Fig. 3.9. Again, the spectra were measured consecutively to provide comparability.[3] Scaling was done with respect to the broad band between 175 and

[3] Other than the He/CF_4 carrier gas mixture which was prepared in a vacuum line, SF_6 and C_2F_6 were mixed to helium in the reservoir preceding the brass saturator. This procedure does not allow to set the concentration accurately. The admixture of SF_6 and C_2F_6 was thus adjusted based on observed band intensities and calculated scattering cross-sections (from harmonic frequency calculations with Gaussian 09 [22] on the B3LYP/6-311++G(d,p) level) to yield concentrations comparable to the helium/tetrafluoromethane mixture (≈4 %).

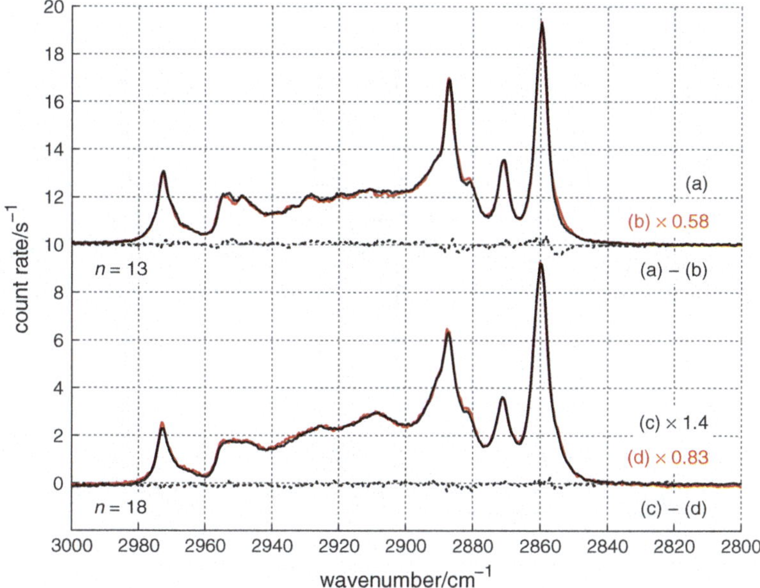

Fig. 3.10 Raman jet spectra of C–H stretching vibrations of tridecane and octadecane expanded in He (**a**, **c**) and He + 4 % CF_4 (**b**, **d**). Difference spectra are plotted with *dotted lines*. The effect of CF_4 addition is seen to be negligible for this spectral region (the higher absolute intensity stems from the increased stagnation pressure). *Measurement conditions* **a** $\vartheta_s = 37\,°C$, $p_0 = 0.5$ bar, $p_b = 0.9$ mbar, $d_n = 1$ mm, exposure 28 × 90 s, **b** $\vartheta_s = 37\,°C$, $p_0 = 0.9$ bar, $p_b = 1.3$ mbar, $d_n = 1$ mm, exposure 24 × 90 s, **c** $\vartheta_s = 85\,°C$, $p_0 = 0.5$ bar, $p_b = 0.8$ mbar, $d_n = 1$ mm, exposure 28 × 180 s, and **d** $\vartheta_s = 90\,°C$, $p_0 = 0.85$ bar, $p_b = 1.2$ mbar, $d_n = 1$ mm, exposure 16 × 60 s

275 cm^{-1}, due to the absence of other prominent gauche signals (the signal at 352 cm^{-1} is a higher order all-trans accordion vibration). The comparison of the spectra is thus a little less reliable than in case of the example discussed above, but a substantial advantage of SF_6 or C_2F_6 over CF_4 can be ruled out: the relative intensity of the accordion vibration band (127 cm^{-1}) does not change when switching from CF_4 to SF_6 or C_2F_6. A somewhat better cooling ability of SF_6 and C_2F_6 was anticipated because of their larger mass, but their higher number of vibrational degrees of freedom (which need to be cooled as well) seem to counterbalance any possible advantage. Therefore, these substances are inferior to CF_4 as carrier gas additive, in so far that they exhibit more complicated Raman spectra which potentially overlap with interesting bands.

For Raman jet spectra of alkanes at higher wavenumbers (C–C-stretching, C–H-bending, and C–H-stretching vibrations), the influence of the carrier gas is almost negligible. In Fig. 3.10, spectra of C–H-stretching vibrations of tridecane and octadecane expanded in helium and helium/tetrafluoromethane are depicted. Residuals after subtracting jet spectra from different carrier gas compositions show that the carrier gas effect is indeed very small. This demonstrates that the C–H-stretching band is insensitive to smaller changes in the conformational distribution. Spectra in the C–C

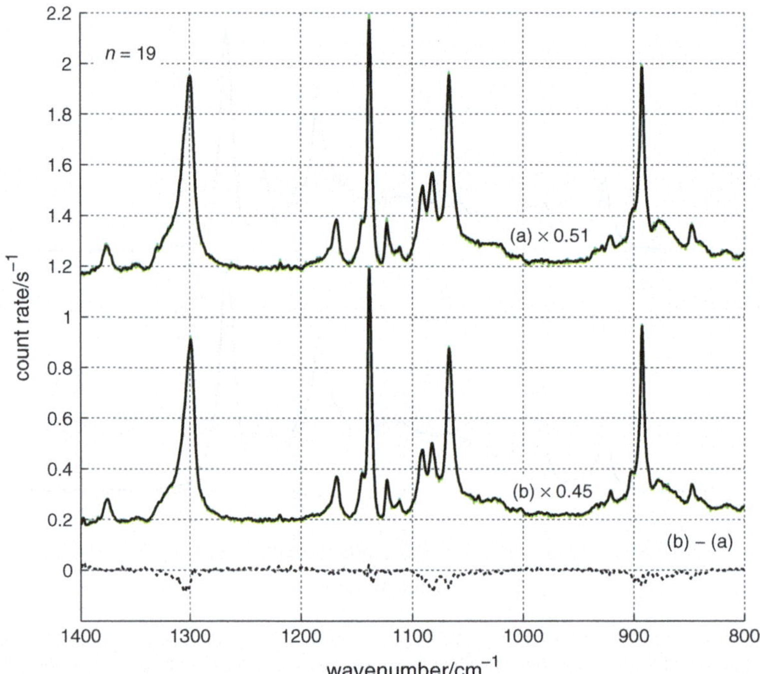

Fig. 3.11 Spectra of the C–C-stretching region of nonadecane in expansions with different carrier gas compositions but otherwise similar experimental parameters. **a** Expansion in He, and **b** expansion in He + 30 % Ne. Original data is plotted in *green*, Savitzky-Golay filtered spectra (9 pt.) in *black*. *Measurement conditions* **a** $\vartheta_s = 105\,°C$, $p_0 = 0.6$ bar, $p_b = 0.8$ mbar, $d_n = 1$ mm, exposure 6×600 s, **b** $\vartheta_s = 105\,°C$, $p_0 = 0.8$ bar, $p_b = 0.8$ mbar, $d_n = 1$ mm, exposure 8×600 s

stretching region were measured in the last stage of this work and not investigated systematically with respect to the carrier gas composition. Effects of adding argon were found to be negligible. Figure 3.11 shows spectra of nonadecane in the C–C stretching region, recorded from helium and helium/neon expansions (30 %). Adding neon leads to a slight decrease of some bands relative to the central sharp band at $\approx 1140\,cm^{-1}$.

Most of the spectra shown so far are Savitzky-Golay filtered for better clarity, using a frame size of 7–9 points and second order polynomials. Therefore, one has to be careful when interpreting weak signals. The octadecane spectra shown in Fig. 3.9 demonstrate the limit of rather low signal-to-noise ratios. Not all of the weak signals are actually reproducible, and should consequently be omitted in the interpretation. On the other hand, the tridecane spectra shown in Fig. 3.8 are examples of higher signal-to-noise ratios, where the filtering is not critical. The spectra which follow in Chaps. 4 and 5 are commonly Savitzky-Golay filtered too, yet plotting of the original data is omitted because the figures would be too congested otherwise. However, reproducibility of weak signals is generally demonstrated by showing spectra of alkanes expanded with different carrier gas mixtures.

References

1. P. Zielke, Ramanstreuung am Überschallstrahl, Wasserstoffbrückendynamik aus neuer Perspektive, Ph.D. Thesis, Georg-August-Universität Göttingen, 2007
2. T.N. Wassermann, Umgebungseinflüsse auf die C-C- und C-O-Torsionsdynamik in Molekülen und Molekülaggregaten: Schwingungsspektroskopie bei tiefen Temperaturen, Ph.D. Thesis, Georg-August-Universität Göttingen, 2009
3. Z. Xue, Raman spectroscopy of carboxylic acid and water aggregates, Ph.D. Thesis, Georg-August-Universität Göttingen, 2010
4. P. Zielke, M.A. Suhm, Concerted proton motion in hydrogen-bonded trimers: a spontaneous Raman scattering perspective. Phys. Chem. Chem. Phys. **8**, 2826–2830 (2006)
5. M. Nedić, T.N. Wassermann, Z. Xue, P. Zielke, M.A. Suhm, Raman spectroscopic evidence for the most stable water/ethanol dimer and for the negative mixing energy in cold water/ethanol trimers. Phys. Chem. Chem. Phys. **10**, 5953–5956 (2008)
6. Z. Xue, M.A. Suhm, Probing the stiffness of the simplest double hydrogen bond: the symmetric hydrogen bond modes of jet-cooled formic acid dimer. J. Chem. Phys. **131**, 054301 (2009)
7. N.O.B. Lüttschwager, T.N. Wassermann, S. Coussan, M.A. Suhm, Periodic bond breaking and making in the electronic ground state on a sub-picosecond timescale: OH bending spectroscopy of malonaldehyde in the frequency domain at low temperature. Phys. Chem. Chem. Phys. **12**, 8201–8207 (2010)
8. S. Hesse, T.N. Wassermann, M.A. Suhm, Brightening and locking a weak and floppy N-H chromophore: the case of pyrrolidine. J. Phys. Chem. A **114**, 10492–10499 (2010)
9. T.N. Wassermann, M.A. Suhm, Ethanol monomers and dimers revisited: a Raman study of conformational preferences and argon nanocoating effects. J. Phys. Chem. A **114**, 8223–8233 (2010)
10. T.N. Wassermann, J. Thelemann, P. Zielke, M.A. Suhm, The stiffness of a fully stretched polyethylene chain: a Raman jet spectroscopy extrapolation. J. Chem. Phys. **131**, 161108 (2009)
11. K. Liu, R.S. Fellers, M.R. Viant, R.P. McLaughlin, M.G. Brown, R.J. Saykally, A long path length pulsed slit valve appropriate for high temperature operation: infrared spectroscopy of jet-cooled large water clusters and nucleotide bases. Rev. Sci. Instrum. **67**, 410–416 (1996)
12. C. Cézard, C.A. Rice, M.A. Suhm, OH-stretching red shifts in bulky hydrogen-bonded alcohols: jet spectroscopy and modeling. J. Phys. Chem. A **110**, 9839–9848 (2006)
13. M. Albrecht, C.A. Rice, M.A. Suhm, Elementary peptide motifs in the gas phase: FTIR aggregation study of formamide, acetamide, *N*-methylformamide, and *N*-methylacetamide. J. Phys. Chem. A **112**, 7530–7542 (2008)
14. M. Goubet, R.A. Motiyenko, F. Réal, L. Margulès, T.R. Huet, P. Asselin, P. Soulard, A. Krasnicki, Z. Kisiel, E.A. Alekseev, Influence of the geometry of a hydrogen bond on conformational stability: a theoretical and experimental study of ethyl carbamate. Phys. Chem. Chem. Phys. **11**, 1719–1728 (2009)
15. R.M. Balabin, Conformational equilibrium in glycine: experimental jet-cooled Raman spectrum. J. Phys. Chem. Lett. **1**, 20–23 (2010)
16. MATLAB, version 7.12.0 (R2011a), The MathWorks Inc., Natick (2011)
17. F. Dietrich, Das Dimer des Acetons, B.Sc. Thesis, Georg-August-Universität Göttingen, 2011
18. *CRC Handbook of Chemistry and Physics*, ed. by D.R. Lide, 82nd edn, (CRC Press, Boca Raton, 2001)
19. R.M. Balabin, Enthalpy difference between conformations of normal alkanes: Raman spectroscopy study of *n*-pentane and *n*-butane. J. Phys. Chem. A **113**, 1012–1019 (2009)
20. S. Bocklitz, Conformational analysis of *n*-alkanes in cryosolutions, M.Sc. Thesis, Georg-August-Universität Göttingen, Universiteit Antwerpen, 2013
21. J. Lee, Adaptive Aggregation über starke Wasserstoffbrücken in der Gasphase, Ph.D. Thesis, Georg-August-Universität Göttingen, 2012

22. M.J. Frisch, G.W. Trucks, H.B. Schlegel, G.E. Scuseria, M.A. Robb, J.R. Cheeseman, G. Scalmani, V. Barone, B. Mennucci, G.A. Petersson, H. Nakatsuji, M. Caricato, X. Li, H.P. Hratchian, A.F. Izmaylov, J. Bloino, G. Zheng, J.L. Sonnenberg, M. Hada, M. Ehara, K. Toyota, R. Fukuda, J. Hasegawa, M. Ishida, T. Nakajima, Y. Honda, O. Kitao, H. Nakai, T. Vreven, J.A. Montgomery, Jr., J.E. Peralta, F. Ogliaro, M. Bearpark, J.J. Heyd, E. Brothers, K.N. Kudin, V.N. Staroverov, R. Kobayashi, J. Normand, K. Raghavachari, A. Rendell, J.C. Burant, S.S. Iyengar, J. Tomasi, M. Cossi, N. Rega, J.M. Millam, M. Klene, J.E. Knox, J.B. Cross, V. Bakken, C. Adamo, J. Jaramillo, R. Gomperts, R.E. Stratmann, O. Yazyev, A.J. Austin, R. Cammi, C. Pomelli, J.W. Ochterski, R.L. Martin, K. Morokuma, V.G. Zakrzewski, G.A. Voth, P. Salvador, J.J. Dannenberg, S. Dapprich, A.D. Daniels, O. Farkas, J.B. Foresman, J.V. Ortiz, J. Cioslowski, D.J. Fox, Gaussian 09, Revision A.02, Gaussian Inc, Wallingford CT (2009)

Chapter 4
Unbranched *n*-Alkanes

In this chapter, the Raman spectroscopic investigation of *n*-alkane self-solvation will be presented. To support the later analysis, the first section discusses the conformational isomerism of *n*-alkanes in general and addresses the question to which temperature alkanes must be cooled to work out low-energy conformations. Section 4.2 outlines the vibrational degrees of freedom of unbranched *n*-alkanes to facilitate the assignment of Raman jet spectra. In Sect. 4.3, the simulation of Raman jet spectra which aid the assignment and provide conformational temperatures is discussed. Raman jet spectra are presented and assigned subsequently in Sect. 4.4. The chapter closes with the determination of the critical folding chain length (n_c) based on the experimental findings and a discussion of computational predictions.

4.1 *n*-Alkane Conformations

The conformational isomerism of unbranched alkanes is governed by the different stable torsional conformations of neighboring carbon-carbon (C–C) bonds. In order to characterize these stable conformations, it is helpful to consider the 2-dimensional potential energy surface (PES) of the C–C torsional angles τ_1 and τ_2 in pentane, which is shown in Fig. 4.1 (from Ref. [1]).[1] The low energy conformations are denoted t for trans ($\tau \approx \pm 180°$) and $g^{\pm}$ for $\pm$ gauche ($\tau \approx \pm 65°$). Higher energy minima involve the "x conformation", that is, a "cross" or perpendicular torsional angle ($\tau \approx \pm 95°$). It occurs when two neighboring C–C segments approach gauche conformations of opposite sign. Such a g^+g^- sequence would bring the two methyl end groups in close vicinity (see Fig. 4.1) and is avoided because of steric repulsion. A "perfect" $+65°/-65°$ g^+g^- sequence does not correspond to a stationary point on the PES but is close to a saddle point ($\tau_1 = -\tau_2 \approx 75°$) connecting two $g^{\pm}x^{\mp}$ conformers.

[1] The shown example is intended to serve an illustrative purpose; more recent calculations of the pentane PES are available (for example Refs. [2–4]).

N.O.B. Lüttschwager, *Raman Spectroscopy of Conformational Rearrangements at Low Temperatures*, Springer Theses, DOI 10.1007/978-3-319-08566-1_4

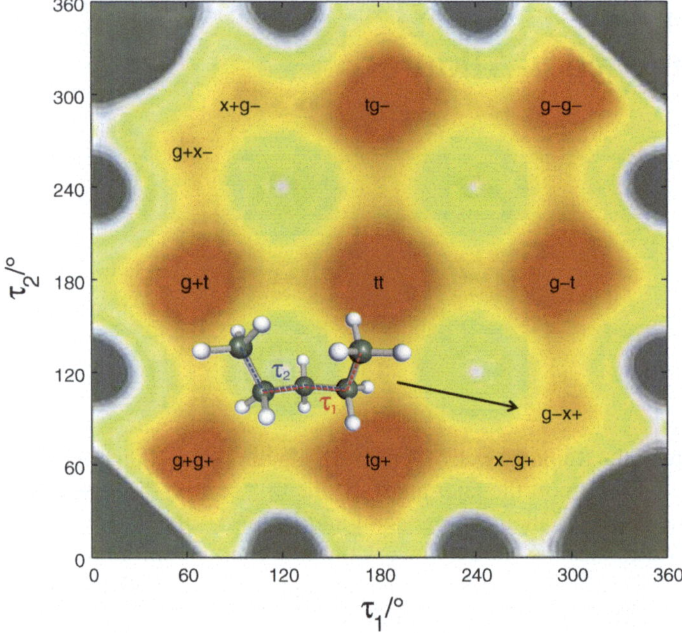

Fig. 4.1 Potential energy surface for the C–C torsions in *n*-pentane. t = trans, $g^{\pm}$ = gauche, $x^{\pm}$ = distorted gauche. Areas of low potential energy are drawn in *red*. The *syn*-pentane conformer $g^{-}x^{+}$ is shown as an example. Reprinted (adapted) with permission from Ref. [1]. Copyright 1998 American Chemical Society

Qualitatively, this potential energy surface is applicable for neighboring C–C bonds in longer alkanes, implying that one has to consider up to five stable conformations for each C–C–C–C torsion angle. Certain sequences are forbidden because they bring chain segments in close vicinity or result in chain overlap [1] and x conformations are restricted to segments directly connected to a gauche sequence. The total number of possible stable conformers N for alkanes with the chain length n should thus fall in the range $3^{n-3} \leq N \leq 5^{n-3}$ [1].[2] Accordingly, the number of possible conformers for alkanes important to this work is of the order 10^6 ($n = 13$) up to 10^{12} ($n = 21$). At first sight, this "haystack" of stable structures seems to render all efforts to reveal the energetic preference between just two conformers hopeless. Indeed, room temperature spectra, either from liquid [5] or gaseous medium sized alkanes [6] are seen to involve many conformers and show no conformer specific signatures. However, the 3^{n-3} to 5^{n-3} counting overestimates the number of structures which are distinguishable by vibrational Raman spectroscopy as it discriminates between enantiomers and conformers of reversed sequence order. Also, the vast majority of the possible conformers are of relatively high energy compared

[2] A hydrocarbon chain with n carbon atoms involves $n-3$ C–C–C–C segments leading to torsional isomerism.

to a small collection of conformers close to the energy minimum [7]. Lowering the temperature by means of free jet expansions was found to allow the observation of single conformer signatures, especially from the all-trans conformer [6, 8].

In order to estimate how the different torsional isomers contribute to the conformational equilibrium of cold isolated alkanes, it is instructive to develop an expression for the conformational partition function. In principle, this demands the knowledge of (1) all stable conformations and (2) the corresponding energies. To address the first problem, one can either systematically count the conformers based on the stable torsional states [1, 9–12] or use random conformational searches such as Monte-Carlo methods [1, 13, 14], molecular dynamics methods [2, 13, 15] or even genetic algorithms [16, 17]. The case of jet-cooled alkanes allows for favorable approximations and the main spectral contributors can be counted in a simple fashion. A first approximation lies in omitting all conformers involving a $g^{\pm}x^{\mp}$ or *syn*-pentane sequence, which are of high energy ($g^{\pm}x^{\mp}$ pentane is approximately 12 kJ mol^{-1} above the all-trans minimum [2, 18]). This approximation is supported by Raman jet spectra of *n*-pentane (Figs. 3.7 and 4.7), low temperature Raman gas phase spectra [19], as well as Raman spectra from pentane in cryosolutions [20], which show that the *syn*-pentane conformer is present in a very small fraction under such conditions. Consistently, one has to omit multi gauche conformers with energies higher than the energy of a *syn*-pentane conformation as well. This leads to the second problem: How much does each gauche conformation contribute to the total energy of a conformer? Several theoretical investigations focused on this question [2, 21–23] and found that the energy is essentially additive provided that the gauche conformations are sufficiently separated by trans segments. Cumulated gauche sequences deviate from additivity as local interactions can have either a stabilizing or destabilizing effect. For example, adjacent gauche torsions of equal rotational sense have a small stabilizing "positive pentane effect" (stabilization by ≈0.7 kJ mol^{-1} [2])[3] because of attractive dispersion interactions between $CH_{2(3)}/CH_{2(3)}$ groups in 1/5 position with respect to the $g^{\pm}g^{\pm}$ sequence [21]. These effects can grow to an important contribution for conformers with a high number of gauche bonds and emphasize the importance of a dispersion correction in DFT calculations (see Sect. 4.3). More details on the gauche energy and its additivity can be found in Appendix A.1. In this work, simple additivity with a uniform gauche penalty will be employed since jet-cooled alkanes will occupy conformations with only a small number of gauche torsions where the probability of finding one or several $g^{\pm}g^{\pm}$ sequences will be rather small.

Assuming additivity of the gauche-energy one can express the conformational partition function q_{conf} in the simple form:

$$q_{\mathrm{conf}} = \frac{1}{G_0} \sum_{i=0}^{n-3} G_i \exp\left(-\frac{i\Delta E_{\mathrm{g,t}}}{RT}\right). \tag{4.1}$$

[3] A prominent example is the case of pentane: The $g^{\pm}g^{\pm}$ conformer is found to have an enthalpy of ≈3.9 kJ mol^{-1} relative to tt—less then twice the enthalpy difference of the single gauche conformer (≈2.6 kJ mol^{-1}) [19].

In this equation, G_i is the (artificial) degeneracy of conformers with i gauche torsions (the number of possible conformations with i gauche bonds), $\Delta E_{\mathrm{g,t}}$ the gauche/trans energy difference or gauche energy penalty, R the ideal gas constant and T the temperature. The normalization factor $1/G_0$ ensures the partition function approaches 1 when $T \rightarrow 0$ K. In the literature, $\Delta E_{\mathrm{g,t}} = 0.5$ kcal mol^{-1} ($\approx$2.1 kJ mol^{-1}) was recommended as a good compromise between higher energy isolated gauche torsions and lower energies from gauche states in a $\mathrm{g}^{\pm}\mathrm{g}^{\pm}$ sequence [2], but also higher values like 0.8 kcal mol^{-1} were reported [24].[4] In this work $\Delta E_{\mathrm{g,t}} = 2.5$ kJ mol^{-1} will be employed with 3.0 kJ mol^{-1} as an upper and 2.0 kJ mol^{-1} as a lower bound. In addition to experimental findings [19, 20, 26] (older values are summarized in Ref. [21]), this value agrees well with recent high level quantum chemical calculations [4, 18]. The reason not to use the value suggested by Klauda et al. [2] (2.1 kJ mol^{-1}) is that conformers with stabilizing $\mathrm{g}^{\pm}\mathrm{g}^{\pm}$ sequences will be rather rare in the low temperature limit, as stated above. In any case, the conformer count model is not intended to precisely predict the distribution of conformers, but rather to allow an estimation of what temperature will be necessary to reach a reasonable small number of populated conformers.

Having established the gauche energy penalty, the coefficients G_i need to be found. Note that disregarding *syn*-pentane sequences is equivalent to rejecting x conformations which reduces the number of possible conformers to $<3^{n-3}$ (3^{n-3} is the upper limit when t and $\mathrm{g}^{\pm}$ is allowed for each torsion regardless of the neighboring conformations). To evaluate and count all 3^{n-3} possible conformers, a Fortran 95 program was written. It runs through a tree diagram [12, 27] as shown in Fig. 4.2 for the case of hexane. In such a diagram, each branch corresponds to one (not necessarily unique) conformer. The program follows each branch and counts the number of gauche conformations but stops and continues with the next branch as soon as it runs into a $\mathrm{g}^{\pm}\mathrm{g}^{\mp}$ sequence. The resulting overall numbers of conformers of this independent code are listed in Table 4.1 and match precisely the results of a mathematical expression developed by Tasi et al. [12]. The source code is outlined in Appendix A.2.

In order to use the conformational model partition function to draw conclusions regarding the conformational distribution in the experiment, one has to assume similar partition functions for the remaining degrees of freedom for all conformers, since these are not accessible in a simple fashion. The symmetry number however, occurring in the rotational partition function, is an exception and should not be ignored. Following an approximate expression of the rotational partition function (more details are given when discussing a related problem in Sect. 4.3), the count coefficients need to be reduced according to the symmetry number. This symmetry issue is reflected in the tree diagram which makes it a convenient way to count the conformers. All-trans can only be realized in one way and occurs as the central branch in the tree

---

[4] Snyder reported this value based on the agreement between simulated and observed Raman spectra of liquid alkanes. The deviation between the energy value used by Snyder and the one used here might stem from different structural preferences of liquid and gaseous alkanes [25] (intermolecular interactions outweigh intramolecular interactions).

Table 4.1 Counted and symmetry weighted number of conformers with i gauche torsions neglecting $g^{\pm}x^{\mp}$ sequences obtained from the Fortran program `count_conf.f95` (see Appendix A.2)

	$n =$	13	14	15	16	17	18	19	20	21
$i = 1$		10	11	12	13	14	15	16	17	18
2		81	100	121	144	169	196	225	256	289
3		344	489	670	891	1156	1469	1834	2255	2736
4		833	1408	2241	3400	4961	7008	9633	12936	17025
5		1182	2471	4712	8361	14002	22363	34332	50973	73542
6		985	2668	6321	13504	26577	48940	85305	142000	227305
7		476	1765	5418	14407	34232	74313	149830	284075	511380
8		129	704	2945	10128	29953	78592	187137	411280	845185
9		18	163	996	4645	17718	57799	166344	432073	1030490
10		1	20	201	1360	7001	29364	104881	329024	927441
11		—	1	22	243	1804	10165	46530	180775	614680
12		—	—	1	24	289	2336	14305	71000	298305
13		—	—	—	1	26	339	2964	19605	104910
14		—	—	—	—	1	28	393	3696	26265
15		—	—	—	—	—	1	30	451	4540
16		—	—	—	—	—	—	1	32	513
17		—	—	—	—	—	—	—	1	34
18		—	—	—	—	—	—	—	—	1
Sum		8119	19601	47321	114243	275807	665857	1607521	3880899	9369319
3^{n-3}		59049	177147	531441	1594323	4782969	14348907	43046721	129140163	387420489

The sum includes $i = 0$ and a factor of 2 (see text)

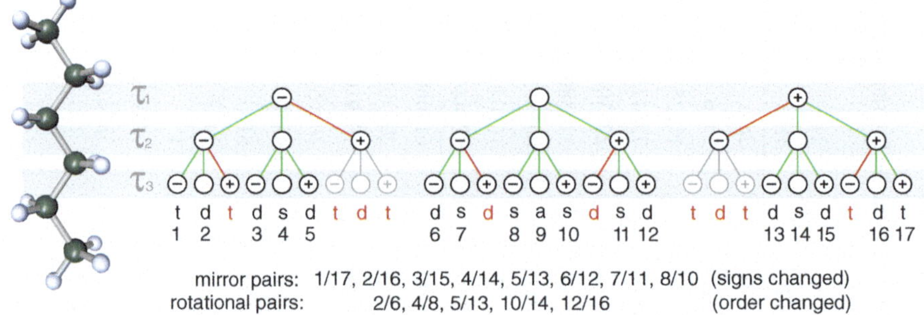

Fig. 4.2 Tree diagram to count hexane conformers. Trans conformations are indicated by *empty circles*, gauche are marked with "+" and "−". Forbidden paths including $g^{\pm}g^{\mp}$ are drawn in *red*, allowed paths in *green*. Conformers are classified as all-trans (a), single gauche (s), double gauche (d), and triple gauche (t). Except for all-trans, each path has an isomorphic partner (conformers are rotational or mirror pairs). Reference I—Copyright Wiley-VCH Verlag GmbH & Co. KGaA. Reproduced with permission

diagram. It has either C_{2v} (odd n) or C_{2h} (even n) symmetry. The remaining branches will contain at least one gauche and each branch has at least one energy equivalent partner with reversed gauche signs (denoted "mirror pairs" in Fig. 4.2), that is, the tree diagram itself (as the alkane PES) is symmetric. Furthermore, the tree diagram contains equivalent branches with reversed sequence order which are denoted "rotational pairs" in Fig. 4.2. An unsymmetric C_1 conformer will occur in four equivalent sequences in the tree diagram, for example the sequences 2/6/12/16 in the hexane case. Some conformers belong to the $C_{2v(h)}$ subgroups C_2 and $C_{s(i)}$. This symmetry property is reflected in the sequence order [9]. For symmetric conformers, reversing the sequence order has either no effect (sequence 1/3/7/11/15/17 in the hexane example, C_2 symmetry [9]) or it is equivalent to reversing all gauche signs (sequences 5/13, C_i symmetry [9]). In this case, the corresponding conformer has only two equivalent branches. To incorporate the symmetry number, the weight of all-trans is reduced to $G_0 = 1/2$. The weight of C_2 symmetric conformers needs to be reduced by 1/2 as well, but they are statistically favored due to two enantiomeric ways of realization. The contribution by one enantiomeric pair to G_i is thus 1. Each enantiomer of a C_1 symmetric conformer is counted as 1. C_s or C_i conformers are not chiral and are counted as 1. These weightings result from the tree diagram without any further evaluation of the conformers symmetry, just by counting all branches as 1/2. Note that this evaluation is valid only in case of pure $^{1}H/^{12}C$ alkanes. Inclusion of other hydrogen or carbon isotopes, which would break the symmetry, is omitted. This limits the accuracy of the conformational partition function for long chains, when the probability of having a rare hydrogen or carbon isotope (especially ^{13}C with a natural abundance of about 1 %) included in the molecule increases.

Limitations regarding the temperature interval in which the model partition function can be applied arise from omitting $g^{\pm}g^{\mp}$ sequences. By means of quantum chemistry, the energy handicap for $g^{\pm}g^{\mp}$ pentane is found to be $\approx$12 kJ mol^{-1} but it decreases slightly with increasing chain length [2, 18]. For conformers with not too many gauche conformations (not too high i), the number of possible conformers with a $g^{\pm}g^{\mp}$ sequence will be small compared to the number of conformers with no such sequence but similar energy (and hence overall more gauche bonds). Therefore, up to a certain number of gauche bonds, neglecting $g^{\pm}g^{\mp}$ will not generate a large error. Yet, it is evident from Table 4.1 that this cannot hold for conformers with a high number of gauche bonds. The number of possible conformers starts to decrease at a certain i, which reflects that there are less and less options to include yet another gauche torsion without creating a $g^{\pm}g^{\mp}$ sequence. This means that at higher i, $g^{\pm}g^{\mp}$ containing conformers cannot be neglected anymore and at some point they will even make up the majority of conformers with a particular (high) i. This is apparent when comparing the sum of all allowed conformers without $g^{\pm}g^{\mp}$ against the number of all 3^{n-3} possible $g^+/g^-/t$ combinations (bottom of Table 4.1), and demonstrates why the bulk of all possible conformers are of rather high energy compared to a smaller set of low energy conformers as stated earlier. The consequence for the partition function is that it should to be truncated at some point rather than include all conformers without $g^{\pm}g^{\mp}$, and one should estimate a temperature interval were it will still be reasonably accurate. Here, the partition function is truncated conservatively after 5 gauche, so that the sum in Eq. 4.1 runs from $i = 0$ to 5. This is about the number where the first neglected $g^{\pm}g^{\mp}$ conformers come close to the considered energy range.

The assessment of this truncated model partition function can be done by calculating its value for different chain lengths and observing how it converges to the value of the "full" partition function (including all coefficients in Table 4.1 but still neglecting $g^{\pm}g^{\mp}$) with decreasing temperature. The result is shown in Fig. 4.3. Because of the increasing number of possible gauche conformers with increasing chain length

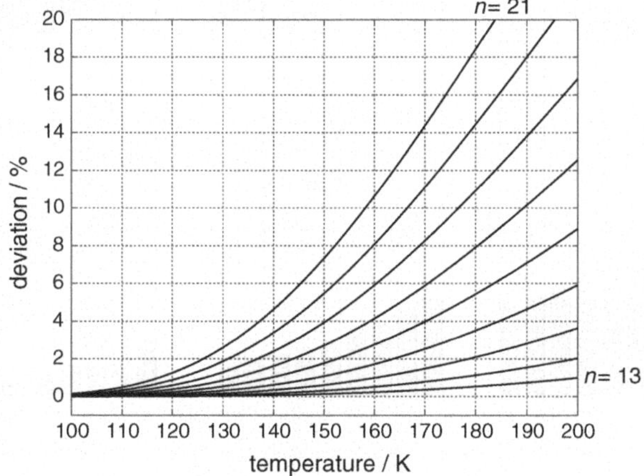

Fig. 4.3 Percentual deviation of the truncated model partition function from the partition function using all coefficients in Table 4.1 with increasing temperature for different chain lengths n

n, the deviation will be highest for longer alkanes. Here the partition function will only be valid for rather low temperatures. In the limit of $n = 21$, deviations of more than 10 % would be expected beyond 160 K. On the other hand, the partition function should be reasonably well suited for temperatures below 150 K for all studied alkanes. The temperature constraint prohibits the direct comparison of this partition function to modelled alkane conformer equilibria, like reported in Ref. [14] for 295 K and 1 atm.

Using the model partition function q_{conf}, the fractions of conformers involving i gauche torsions can be calculated by:

$$\frac{N_i}{N} = \frac{1}{G_0} \times G_i \exp\left(-\frac{i\Delta E_{\text{g,t}}}{RT}\right) \times \frac{1}{q_{\text{conf}}}.$$

The all-trans fraction N_0/N is:

$$\frac{N_0}{N} = \frac{1}{q_{\text{conf}}}.$$

The results for $n = 13$–21 are shown in Fig. 4.4 in the form of bar diagrams where the different colors indicate different quantities of gauche torsions per chain, and the height of each bar corresponds to the conformer fraction. The fractions are calculated employing the gauche energy penalty $\Delta E_{\text{g,t}} = 2.5$ kJ mol^{-1}, but the graphs can be translated to the upper and lower bounds (2.0 and 3.0 kJ mol^{-1}) since only the ratio of the energy difference to the temperature $\Delta E_{\text{g,t}}/T$ enters the underlying

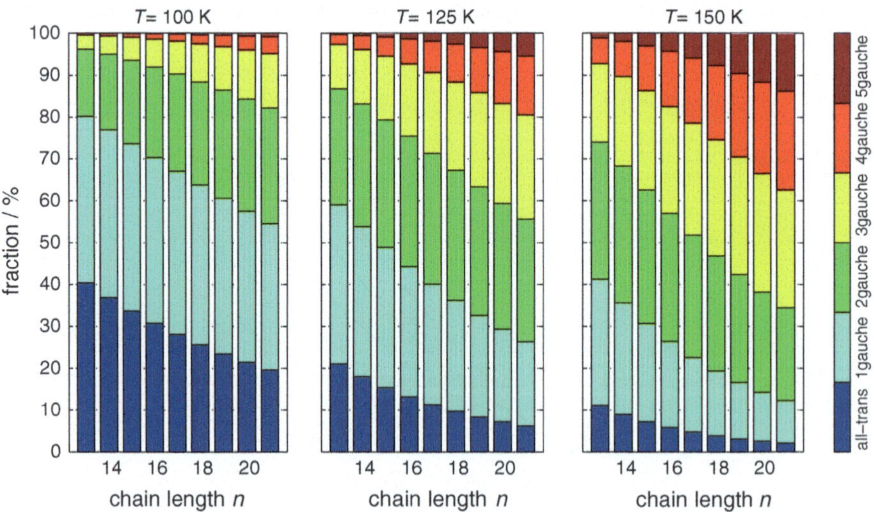

Fig. 4.4 Conformer fractions calculated for different chain lengths and temperatures using the model partition function (Eq. 4.1) with $\Delta E_{\text{g,t}} = 2.5$ kJ mol^{-1}. Reference II—Reproduced by permission of The Royal Society of Chemistry

equations. Changing the energy to 2.0 (3.0) kJ mol^{-1} would change the temperatures in Fig. 4.4 from left to right to 80 (120), 100 (150) and 120 (180) K, respectively. At a conformational temperature of 100 K, the majority of conformers are single gauche and all-trans. While the fraction of single gauche conformers decreases only slightly with increasing chain length, the all-trans fraction drops by ≈1/2 from $n = 13$ to 21. The fractions of conformers with more than one gauche torsion increase with n, but conformers with $i = 4$ and $i = 5$ stay rather negligible. At higher temperatures, the statistically favored conformers with $i > 1$ start to outweigh all-trans and single gauche. Whether a group of conformers is preferred or handicapped at higher chain lengths depends on the temperature. For example, the fraction of double gauche conformers increases with increasing n at 100 K, stays rather constant at 125 K and finally decreases at 150 K. At this temperature, the highest all-trans contribution is ≈10 % for tridecane ($n = 13$). At higher chain length, the all-trans fraction only makes up a few percent. Therefore, judging from the model partition function, temperatures below 150 K are beneficial, in order not to obscure energetically favored hairpin conformers by the high amount of multiple gauche conformers. Also at 150 K and $n = 21$ the model is seen to surpass its limits, as conformers with five gauche torsions start to make an important contribution.

Beside inaccuracies due to approximations, one has to keep in mind that this estimation of conformer fractions at different temperatures applies for molecule ensembles in thermal equilibrium. Jet-isolation involves rapid cooling so that higher gauche conformers can freeze in their high energy state and the final conformer distribution will deviate from thermal equilibrium. However, the conformational partition function was used to estimate an overall effective conformational temperature and it will be seen that this value does not drastically deviate from a different estimation based on the fraction of single gauche conformers relative to all-trans conformers (Sect. 4.5.1). The freezing of conformational states should thus be rather uniform and the partition function reasonably valid, at least for the single gauche to trans conversion.

4.2 n-Alkane Vibrations Relevant to this Work

Before dealing with the spectral assignment, some general comments on how hydrocarbon chains vibrate shall be made. This facilitates the comparison to the literature and helps classifying different types of vibrations. For this purpose, the polyethylene dispersion curves are especially helpful and will be introduced. This immediately simplifies the description of the numerous vibrational degrees of freedom which alkanes in the considered size range possess (117 for tridecane with $n = 13$ and 189 for heneicosane with $n = 21$).[5]

[5] The number of vibrational degrees of freedom f_{vib} may be calculated by $f_{\text{vib}} = 3N - 6$, using the overall number of atoms N, or, more conveniently, by $f_{\text{vib}} = 9n$, using the chain length n ($N = 3n + 2$).

A first approach to describe the vibrations of alkanes starts with noticing that they behave like systems of coupled identical oscillators, the methylene groups. A qualitatively correct picture of normal vibrations is then available from the textbook example of identical coupled point masses oscillating in one dimension [28]. Solving the equations of motion for such a simplified system containing N point masses yields N normal modes which describe N sinusoidal standing waves. These standing waves differ in their number of nodes which runs from 0, a complete in-phase movement equivalent to a zero-frequency translation, to $N - 1$, a complete out-of-phase vibration. Another way to characterize these normal modes, related to the number of nodes, is using a phase difference (φ) which relates amplitudes (A) of adjacent oscillators, such that

$$A_{n,k}(\varphi) \propto \cos(\varphi_k n) \tag{4.2}$$

where n is the position of the oscillator in the chain. The phase can assume values $\varphi_k = \pi k/N$ with the integer $k = 0, 1, \ldots, N-1$. k may then be used for reference to a normal mode [29]. With $k = 0$, the amplitudes all assume the same value describing a complete in-phase movement, while a complete out-of-phase vibration results from the highest k value $N - 1$. For the limiting case of an infinite number of coupled oscillators, the phase can assume an arbitrary value between 0 and π. Plotting the associated normal mode frequencies against the phase yields a frequency branch or "dispersion curve", which is common in solid state physics to describe the vibrations of crystal lattices [30]. The connection to alkanes is realized when thinking of an isolated hydrocarbon chain as a one dimensional crystal.

To apply the coupled oscillator model to linear n-alkanes one needs to identify appropriate internal coordinates. Looking at an isolated methylene unit, its internal coordinates are the C–H distances and the H–C–H angle. A normal coordinate analysis for this system would yield a symmetric and antisymmetric stretching vibration as well as a bending vibration (also denoted scissoring vibration), much like in the related case of the water molecule. For methylene units lined up in a linear zig-zag chain, further vibrational degrees of freedom arise from the translational and rotational degrees of freedom of the isolated methylene unit, which are hindered for a subunit in the chain. They translate to C–C stretching, C–C–C bending, and C–C–C–C torsion, as well as methylene rocking, twisting, and wagging coordinates. Figure 4.5 illustrates the methylene deformation coordinates. Treating these internal coordinates as separate sets of coupled oscillators would yield normal modes analogous to the simple case of point masses in one dimension discussed above, characterized by a phase or number of nodes. However, the various internal coordinates couple to each other and the situation is more complicated. Yet, if the methylene chain is long, coupling of internal coordinates with different phase values will be small [31] and the picture of normal modes which are characterized by a phase or a certain number of nodes remains valid.

The limiting case of a virtually infinite number of coupled methylene units lined up on an all-trans chain is closely realized by crystalline polyethylene [32]. The vibrational problem can be approached by a clever choice of internal coordinates which

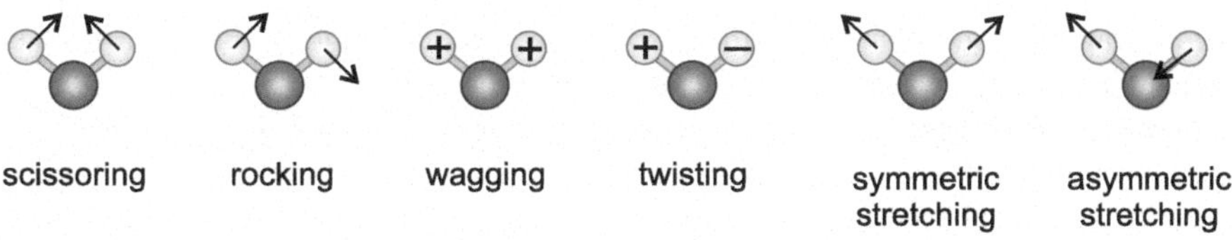

Fig. 4.5 Different types of CH_2 deformation motions; wagging and twisting displacement is perpendicular to the paper plane

relate displacements of identical subunits by a phase factor, following the results of the coupled oscillator approach. Solving the equations of motion for each phase value [32, 33] yields nine frequency branches, originating from the nine degrees of freedom inherent to each methylene unit. These frequency branches can be analyzed with respect to the contributing internal coordinates which gives a vibrational classification for each branch [31]. The two lowest-frequency branches are two acoustical branches (the frequency approaches zero for $\varphi = 0$ or π): The C–C–C bending branch, which is characterized by longitudinal stretching and transversal in-plane bending, and the C–C–C–C torsion branch, which is characterized by twisting and transversal out-of-plane bending of the carbon chain [34]. The remaining seven branches are optical branches with a non-zero frequency for the limiting phase values. From low to high frequency, the vibrational displacement corresponds mainly to CH_2 twisting/rocking, C–C stretching, CH_2 rocking/twisting, CH_2 wagging, CH_2 scissoring, and symmetric and antisymmetric CH_2 stretching. The internal coordinate contribution for one frequency branch may vary with the phase value [31]. The twisting/rocking branch for example is denoted this way because the internal coordinate contribution changes from pure twisting to pure rocking at the limiting phase values (the opposite internal coordinate contribution is found for the rocking/twisting branch) [31].

The analogy to polyethylene is convenient, because normal modes of n-alkanes lie at finite phase values on the polyethylene dispersion curves [34] if neither the coupling to methyl vibrations is large nor the coupling of coordinates with different phases is large. Thus, the main part of the $9n$ normal vibrations of an n-alkane molecule can be identified by their affiliation to a polyethylene frequency branch and a phase index k [29]. This work will classify vibrations by the number of nodes m, because k is not uniformly assigned in the literature (for example the most in-phase C–C stretching vibration in Refs. [29] and [35]). Vibrations which are necessarily not covered by the polyethylene frequency branches are the end group methyl vibrations. They can be identified by their relative intensity and constant wavenumber.

For all-trans methylene chains, the phase of a mode defines its symmetry and optical activity. For polyethylene, only modes with $\varphi = 0$ or π are Raman or IR active, because local transition moments cancel for all other phase values [32]. This is not the case for finite n-alkanes, and consequently, one observes band progressions for vibrations belonging to the same branch but differing in the phase. However, the intensities of vibrations quickly decrease if the phase deviates from 0 or π, because overall transition moments are still small in this case. Vibrations with high Raman intensity are totally symmetric "most in-phase" vibrations.

When conformational disorder is considered, the vibrational problem becomes more complicated [32, 36], and vibrational spectra become more complex. Nevertheless, regularity and band progressions are still observed for spectral series of conformationally disordered alkanes [5, 35], because some vibrations are rather insensitive to the introduction of one or even a few gauche bonds [35]. Single gauche bonds and gauche sequences also give rise to localized modes with characteristic frequencies which can be exploited to analyze the conformation of alkyl chains in more complex systems [37].

Modes of particular interest to this work need to be (1) sufficiently Raman active, and (2) sensitive to folding. Modes which meet these criteria and were investigated systematically belong to the low-frequency acoustical C–C–C bending branch, optical C–C stretching branch, and optical C–H stretching branch.

One of the modes studied in detail is the highly Raman active all-trans accordion vibration belonging to the low-frequency C–C–C bending branch. It is also called longitudinal acoustic mode with one node (LAM-1) because it is characterized by longitudinal stretching of the all-trans chain with a node in the center (see Fig. 4.6), and the associated frequency branch of polyethylene is an acoustical one [31]. Further modes of the same frequency branch give rise to LAMs of higher order with quickly decreasing intensity. The Raman activity of the LAMs is dictated by symmetry. For alkanes with an even chain length n, the symmetry of the all-trans conformer is C_{2h}. LAMs with an even number of nodes are of b_u symmetry and Raman-inactive, those with an odd number of nodes are of a_g symmetry and Raman-active [31]. The opposite is true in case of IR absorption spectroscopy, because the rule of mutual exclusion applies. For odd n the all-trans conformer symmetry is C_{2v}. In this case, LAMs with an even number of nodes (b_1 symmetry) are not strictly forbidden, but still rather weak Raman scatterers.

The fact that accordion-like skeletal vibrations must be sensitive to conformation becomes immediately evident when comparing vibrational Raman spectra of solid and liquid state alkanes in the low-frequency region. Formerly sharp signals will smoothen and make way for the very broad and asymmetric "D-LAM" band (LAM indicating conformational disorder [38, 39]) centered around 200 cm^{-1} [40] when going from ordered crystalline alkanes in the all-trans configuration [41, 42] to disordered gauche conformers in the melt [38, 40]. Looking more closely at specific conformations, one finds that upon breaking the symmetry of the all-trans system by introducing gauche torsions, several vibrations share character of the accordion vibration and become Raman active. Therefore, gauche conformers often contribute with more than one intense band to the low-frequency Raman spectrum. As will will be shown in Sect. 4.4.1, an important consequence for hairpin conformers is the particularly successful detection in the low-frequency region, because of redundant evidence. Less often, redistribution of intensity occurs in case of the all-trans conformation due to harmonic mode mixing or Fermi resonance. These phenomena will be discussed in Chap. 6, where mechanical properties of methylene chains are deduced from the accordion vibration wavenumber [8, 42, 43].

Further Raman-active modes of interest are C–C stretching vibrations between 800 and 1100 cm^{-1}. Conformational disorder affects the wavenumbers of C–C stretching vibrations less than the low-frequency C–C–C bending modes and a systematic pattern remains [5], such that modes indicative of gauche conformers accumulate at certain positions and give rise to rather sharp bands. It will be shown that the hairpin detection succeeds because the $m = 0$ and $m = 2$ C–C stretching vibrations of the hairpin conformer are blue-shifted from their all-trans and single gauche counterparts. Detection based on C–C stretching "defect modes" (localized mainly in specific conformational defects) was suggested to indicate the presence of sharp kink ggtgg sequences by Zerbi and Gussoni [37], but fails close to the most in-phase C–C stretching mode due to spectral congestion. However, a ggtgg defect mode at $\approx$890 cm^{-1} occurs at high hairpin abundance, in line with predictions and experimental evidence from Ref. [37].

Spectra of the C–H stretching band at $\approx$2900 cm^{-1} do not provide substantial hairpin evidence on their own but show some interesting developments in accordance with hairpin evidence from lower frequency spectra. The most in-phase $m = 0$ symmetric and antisymmetric methylene stretching vibrations give rise to the most prominent bands, accompanied by two symmetric and asymmetric methyl stretching vibration bands.

Some of the relevant vibrations are illustrated in Fig. 4.6, calculated on the B3LYP-D3/6-311++G** level using Turbomole v6.4 [44] (Sect. 4.3) and the example of all-trans heptadecane which falls right in the center of the considered size range. Ball-and-stick normal mode illustrations are accompanied by stem plots describing the associated relative amplitudes of C–C stretching and C–C–C bending internal coordinates. These amplitudes were calculated from Cartesian displacement vectors provided by Turbomole (in the limit of small displacements, since the relation between Cartesian coordinates and internal coordinates is approximately linear only in this case). It is emphasized that all stem plots show relative displacements, not absolute displacements, and comparison of amplitudes corresponding to different normal vibrations is not meaningful. This is also the reason why the y-axis is omitted.

Stem plots benefit from a less crowded picture of normal mode displacements and are used in Sect. 4.4.2 to illustrate C–C stretching modes, where more complicated gauche conformer normal coordinates demand a simplified picture. However, the internal coordinate picture is also helpful in case of the all-trans conformation. The C–C stretching vibration of the end groups (center in Fig. 4.6) demonstrates how internal coordinate displacements help to identify localization of normal modes. The LAM-1 is seen to differ from the $m = 0$ C–C stretching vibration by the opposite phase relation of C–C stretching to C–C–C bending. Characterization of phase relations and identification of m values is much simplified when considering the internal coordinate illustration in general.

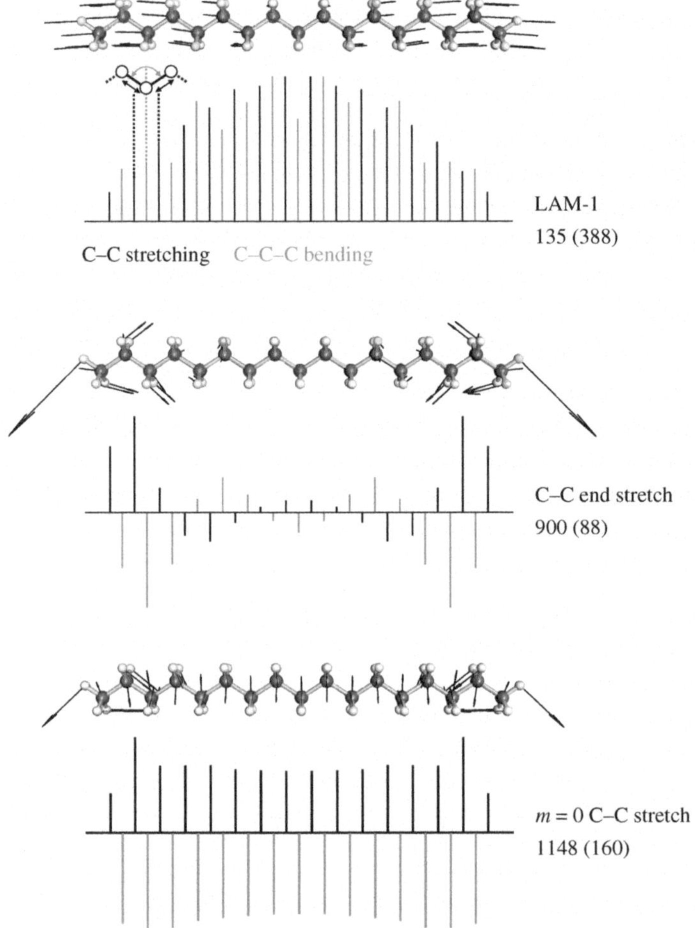

Fig. 4.6 Harmonic normal coordinates of some of the relevant alkane vibrations with major contribution to the Raman spectrum, calculated with Turbomole v6.4 [44] on the B3LYP-D3/6-311++G** level. Wavenumbers in cm^{-1} and the differential scattering cross-section (in parenthesis, $T_{\mathrm{vib}} = 100$ K) in $10^{-36}\ \mathrm{m}^2\,\mathrm{sr}^{-1}$ are given on the *right side*. Relative internal coordinate displacements are drawn below the molecular structures and aligned so that C–C stretching (*black*) and C–C–C bending displacements (*gray*) match the corresponding C–C bonds. The heptadecane all-trans conformer was chosen as a representative chain length of medium size

4.3 Simulation of Raman Jet Spectra

In this section, the simulation of Raman spectra based on quantum chemistry is outlined. Later sections will show that simulated spectra provide a solid ground to assign a majority of the bands observed in experimental Raman jet spectra. Other than spectra of liquid alkanes [5], spectra of jet-cooled alkanes can be approached by computational methods beyond molecular mechanics [45], because, apart from hairpin conformers, conformers with two or more gauche conformations do not contribute significantly, and calculations can be confined to all-trans, hairpin, and single gauche conformers. For the whole set of chain lengths considered in this work ($n = 13$–21), these are less than 100 conformers. The simulation of Raman spectra can be grouped in mainly four steps: (1) Structure optimization and subsequent vibrational analysis using a quantum chemistry software suite, (2) calculation of Raman scattering cross-sections, (3) calculation of vibrational spectra of single conformers using Gaussian bell functions to model vibrational bands, and (4) weighting and summation of single conformer spectra to yield the final simulated Raman spectrum. These steps shall be discussed in the following paragraphs. Afterwards, the accuracy of the prediction will be discussed and some examples will be shown.

4.3.1 Quantum Chemical Approach and Software

The quantum chemical method chosen for structure optimization and vibrational analysis is the hybrid density functional theory approach B3LYP [46–49]. B3LYP is modest regarding computational costs but yet very strong in predicting vibrational wavenumbers of hydrocarbons [8]. A drawback of B3LYP is the insufficient description of dispersion interactions [50] which are mandatory to yield a valid structure for the hairpin conformer. This problem can be tackled by adding a "dispersion correction" [51–53] to the B3LYP energy. For this purpose, the 2010 version of the "D3" correction by Grimme [53] was used. D3 treats dispersion interaction by adding an atom-pairwise term (and a smaller three-body term) to the potential energy of the molecule. The combination of B3LYP and D3 (B3LYP-D3) is available within the software package Turbomole v6.4 [44], which was used for structure optimization, calculation of vibrational frequencies, and calculation of polarizability tensor invariants from which Raman scattering cross-sections are computed. The calculations were carried out using the Pople basis set 6-311++G** [54, 55], the simpler basis set def-SV(P) [56, 57] (in combination with the RI-approximation [58]), and the computationally more expensive basis set def2-TZVP [59]. Because of the high computational effort, def2-TZVP was applied only to all-trans and hairpin conformers. In each case, the energy convergence criterion was set to at least 10^{-7} hartree, the gradient norm convergence criterion to at least 10^{-4} hartree bohr^{-1}. The m4 grid [60] was used throughout. Spectra derived from calculations using the rather limited def-SV(P) basis set were found to be only slightly inferior to spectra derived from calculations using the Pople basis set. Beyond that, however, no systematic search for an optimal basis set was carried out. The discussion will be restricted to calculations employing the basis sets 6-311++G** (spectra, this section and Sect. 4.4) and def2-TZVP (energy, Sect. 4.5).

4.3.2 Single Conformer Spectra

Frequency calculations were carried out within the double harmonic approximation,[6] so that predicted harmonic vibrational wavenumbers are usually too large compared to anharmonic experimental values. To correct for this issue of the harmonic approximation, wavenumbers are scaled [61, 62] by 0.98 or 0.99 to fit the experimental values. Raman scattering cross-sections (Sect. 2.1) are computed from vibrational wavenumbers and polarizability tensor invariants which are available from the Turbomole program output. On top of the uncertainty in the quantum chemically predicted quantities, the vibrational temperature occurring in the equation of the scattering cross-section may be poorly defined. In case of jet-isolated molecules, it might differ for each vibration (Sect. 2.2), and an *effective* vibrational temperature can only be estimated. The value 100 K was used in the calculation of the scattering cross-section throughout, which is about the average temperature derived from the spectral contribution of single gauche conformers, compared to contributions from the all-trans conformer. Scattering cross-sections provide peak intensities of Gaussian curves which are used to model vibrational bands. The exact form of the applied Gaussian function is shown in more detail in Appendix A.3. The width of the Gaussian curves is not calculated but taken from the average FWHM (full width at half maximum) of the accordion vibration band (about 4 cm^{-1} for $n = 13$–21). For all-trans and single gauche conformers, the width is set uniformly to this value. In case of hairpin conformers, the experimental bands are broader and the width of simulated bands is varied accordingly. The integral of a simulated band is proportional to the product of its scattering cross-section and width.

4.3.3 Weighting and Multi Conformer Spectra

Single conformer spectra need to be weighted and merged to yield a final simulated Raman spectrum. Calculating an appropriate weighting factor is challenging for several reasons. The first problem is to capture the conformational distribution of a non-equilibrium jet-cooled alkane ensemble by a physical model. A sophisticated approach would need to describe correctly the kinetics of the conformational isomerizations taking place during the expansion, involving the knowledge of the exact properties of the expansion and all rate constants. This is far outside the scope of this work. However, single gauche and all-trans conformers are separated by just one torsional isomerization of uniform barrier height, so that one may assume the isomerization reactions to be fast enough to yield a conformational distribution close to

[6] In the double harmonic approximation, the force pulling the molecule back to its equilibrium geometry is proportional to the displacement along the normal coordinates, yielding a parabolic potential. On top of this, the polarizability is treated as depending linearly on the displacement along the normal coordinates. Terms of higher order in the displacement are neglected for both, the force and the polarizability.

thermal equilibrium or at least a uniform freezing temperature at which isomerization ceases to be feasible. This means that one single *effective* conformational temperature defines the abundance of the different single gauche conformers relative to the all-trans conformer. In this case, Boltzmann statistics can be applied. On the other hand, for multi gauche conformers, such as the hairpin conformer, which are separated from all-trans by several torsional isomerizations, Boltzmann statistics with this reference temperature are likely to fail, because kinetics play a more important role and barrier heights vary. Because of this, the hairpin conformer fraction is not taken from a Boltzmann weighting but chosen to fit the experimental spectrum. More discussion of this aspect will follow in Sect. 4.5. The manual weighting provides a semi-empirical estimate of the all-trans/hairpin abundance ratio.

A further problem is the correct prediction of the energy difference of conformers. This is a question of suitable electronic structure theory and suitable description of molecular vibrations which contribute with zero-point vibrational energy. Thermal excitation modifies the energy difference predicted in the low temperature limit. A correct prediction of this effect is complicated, since the underlying calculations involve some substantial approximations (see below) and are very sensitive to quantum chemical inaccuracies.

Weighting for a single gauche and all-trans conformer in thermal equilibrium can be performed using the equations of statistical thermodynamics as shown in the following. For a general isomerization reaction:

$$\text{conformer } i \rightleftharpoons \text{conformer } j,$$

the equilibrium constant (K_{ij}) gives the abundance ratio of conformer j to conformer i $\left(\frac{N_j}{N_i}\right)$. It is connected to the Gibbs energy difference at standard conditions (ΔG°_{ij}) by:

$$\Delta G^\circ_{ij} = -RT \ln K_{ij},$$

which can be rearranged to:

$$K_{ij} = \frac{N_j}{N_i} = \exp\left(-\frac{\Delta G^\circ_{ij}}{RT}\right). \tag{4.3}$$

In Appendix A.4, the connection between ΔG°_{ij} and the molecular partition functions is derived, showing that for a conformational isomerization one finds:

$$\Delta G_{ij} = \Delta E^0_{ij} - RT \ln\left(\frac{g_j\, q_{j,\text{rot}}\, q_{j,\text{vib}}}{g_i\, q_{i,\text{rot}}\, q_{i,\text{vib}}}\right) \tag{4.4}$$

where $q_{\text{rot/vib}}$ are rotational or vibrational molecular partition functions and E^0_{ij} is the electronic and zero-point vibrational energy difference (the energy difference at absolute zero). Weighting factors (g) account for enantiomeric degeneracy.

Equation 4.4 is an approximation for non-interacting particles (ideal gas) and non-interacting rotational and vibrational states (no coupling). Combination with Eq. 4.3 and rearrangement leads to:

$$\frac{N_j}{N_i} = \frac{g_j\, q_{j,\mathrm{rot}}\, q_{j,\mathrm{vib}}}{g_i\, q_{i,\mathrm{rot}}\, q_{i,\mathrm{vib}}} \cdot \exp\left(-\frac{\Delta E^0_{ij}}{RT}\right). \tag{4.5}$$

In this equation, all quantities intrinsic to the conformers are available from quantum chemical calculations with Turbomole. The energy difference at a hypothetical temperature of 0 K involves electronic energies from self-consistent field calculations as well as zero-point vibrational energies from harmonic frequency calculations. The rotational molecular partition function is available after geometry optimization and subsequent calculation of the rotational constants. As the zero-point vibrational energy, the vibrational partition function is available from calculated vibrational wavenumbers.

Both molecular partition functions can be computed by the program "freeh", which is a part of the Turbomole software package. freeh calculates the rotational partition function assuming a quasi-continuum of rotational states, which are usually very closely spaced for large molecules (spacing much smaller than $k_B T$). The summation occurring in the general expression of the partition function is then replaced by an integral, and the molecular rotational partition function can be written as [63]:

$$q_{\mathrm{rot}}(T) = \frac{\sqrt{\pi I_A I_B I_C}}{\sigma} \cdot \left(\frac{8\pi^2 k_B T}{h^2}\right)^{\frac{3}{2}}, \tag{4.6}$$

where I are moments of inertia along the molecules principle axes and σ is the symmetry number. The all-trans conformer involves the smallest moment of inertia (and thus the largest rotational constant $A = h/8\pi^2 c I_A$), but even in this case, the quasi-continuum approximation is valid.[7] The vibrational partition function is calculated applying the harmonic approximation which leads to a simple expression [63]:

$$q_{\mathrm{vib}}(T) = \prod_{k=1}^{\text{normal vibrations}} \left(\frac{1}{1-\exp(-hc\tilde{\nu}_k/k_B T)}\right) \tag{4.7}$$

In this equation, the product runs over all normal vibrations and $\tilde{\nu}_k$ is the wavenumber of the kth normal vibration. Both, rotational and vibrational partition functions depend on the temperature and incorporate thermal excitation.

The rotational partition function depends only on the molecular geometry and is not very sensitive to quantum chemical inaccuracies. On the contrary, the vibrational

[7] For the all-trans conformer of the shortest alkane considered, tridecane ($n = 13$), the largest rotational constant is $\approx 0.21\ \mathrm{cm}^{-1}$. Calculating for a linear rigid rotator, this would translate to a characteristic rotational temperature $\theta_{\mathrm{rot}} < 0.5$ K. The classical approximation holds for $T \gg \theta_{\mathrm{rot}}$, and thus in this case even at low rotational temperatures in jet expansions.

partition function is very sensitive to errors in the quantum chemical vibrational analysis. Unbranched alkanes perform very soft skeletal vibrations at low wavenumbers, which make up the main part of the vibrational partition function. These vibrations may vary a lot from the harmonic oscillator description and their prediction by quantum chemistry is often not reliable because they are very sensitive to small errors in the potential energy hypersurface. Indeed, vibrational analysis with Turbomole yielded small imaginary wavenumbers for some conformers of longer alkanes ($n \geq 18$). However, not more than one imaginary wavenumber for each conformer with values not smaller than $-15\,\text{cm}^{-1}$ was found, indicating numerical inaccuracies (rather than the presence of true transition states) as the source of this problem. Even if no imaginary wavenumber occurs, small wavenumber vibrations may cause large uncertainties in the vibrational partition function. A good example is the all-trans conformer of nonadecane ($n = 19$) with the smallest wavenumber being $2\,\text{cm}^{-1}$, calculated at the B3LYP-D3/6-311++G** level. At such low wavenumbers, small errors have a tremendous effect on the vibrational partition function, even at very low temperatures.[8]

As a consequence of this sensitive wavenumber dependence, vibrational partition functions are dropped from calculations of single gauche to all-trans abundance ratios. Necessarily, similar vibrational partition functions are assumed for all-trans and single gauche conformers, and the ratio $\frac{q_{j,\text{vib}}}{q_{i,\text{vib}}}$ in Eq. 4.5 is set to one. This extends the uncertainty of the Gibbs energy difference, which is flawed by inaccuracies of the absolute zero energy difference in any case. Including the full rotational molecular partition functions will not improve the accuracy of the weighting significantly, and differences beside the symmetry number of molecules are omitted as well. This leads to a simpler weighting equation:

$$\frac{N_j}{N_i} = \frac{g_j}{g_i} \cdot \frac{\sigma_i}{\sigma_j} \cdot \exp\left(-\frac{\Delta E_{ij}^0}{RT}\right) \tag{4.8}$$

where populated rotational levels of all conformers are approximated as having equal energies but some are reduced in weight according to the symmetry number. The factors g/σ occurring in this equation are straightforwardly identified. The all-trans conformer is either of C_{2v} or C_{2h} symmetry and is therefore not chiral, leading to $g_\text{t}/\sigma_\text{t} = 1/2$. Single gauche conformers occur in enantiomeric pairs. They are unsymmetric ($g_\text{g}/\sigma_\text{g} = 2/1 = 2$) with one exception: in case of alkanes with an even chain length, the chain center divides a C–C bond. If this bond is twisted into a gauche conformation, the resulting conformer is of C_2 symmetry, and its weight is reduced accordingly ($g_\text{g}/\sigma_\text{g} = 2/2 = 1$).

In Tables 4.3 and 4.2, abundance ratios calculated using Eq. 4.5 and the more simplified Eq. 4.8 are compared, using the example of pentane, hexane, and tridecane. In these cases, the variation is very mild, and vibrational and rotational thermal contributions often cancel. For longer alkanes, when the vibrational partition functions

[8] This is also apparent from the fact that vibrational partition function approaches infinity, in case a wavenumber approaches zero (see Eq. 4.7).

Table 4.2 Single gauche to all-trans abundance in percent from Eqs. 4.5 and 4.8, calculated for tridecane at $T = 110$ K on the B3LYP-D3/6-311++G** level (the ratio of rotational partition functions is temperature-independent)

Position	$q_{g,rot}/q_{t,rot}$	$q_{g,vib}/q_{t,vib}$	ΔE^0_{tg}	Equation 4.5 (%)	Equation 4.8 (%)
1	2.23	0.91	2.94	16	16
2	2.48	0.92	2.74	20	22
3	2.84	0.82	2.64	22	25
4	3.05	0.69	2.72	21	21
5	3.19	0.75	2.62	23	27

The first column refers to the position of the gauche conformation in the alkane chain (see also Fig. 4.9). $g_g/g_t = 2$ and $\sigma_t/\sigma_g = 2$ applies for all single gauche conformers in case of tridecane. ΔE^0_{tg} is the energy difference at 0 K in kJ mol^{-1} (electronic energy plus ZPVE)

starts to suffer from faulty predictions of low-frequency vibrations, the deviation is much more pronounced. Single gauche energies and weightings resulting from Eq. 4.8, used in the simulation of low-frequency spectra for $n = 13$–21, are listed in Table 4.4.

One can see from Table 4.3 that conformational energies of small alkanes predicted on the B3LYP-D3/6-311++G** level agree perfectly with recent high level *ab initio* extrapolations from Gruzman et al. [18], which are included as reference. The values deviate by only 1–6 %. It is worth noticing that this superb agreement depends strongly on the inclusion of Grimme's dispersion correction [53]. This is shown in more detail in Appendix A.1. Single gauche energies of longer alkanes lie all in the reasonable range 2–3 kJ mol^{-1}. Using Grimme's dispersion correction, inaccuracies from the B3LYP energy predictions (which is a point of discussion in its own right [18, 50, 64]) should be minor.

4.3.4 Accuracy of Estimated Conformer Fractions

The accuracy of conformer fractions which can be estimated by comparing simulated and experimental jet spectra (or of the related conformational temperatures) is particularly important to the later analysis and needs discussion. It is linked to the accuracy of the quantum chemically predicted polarizability derivatives (Raman activities), the vibrational temperature which enters the scattering cross-section, and the visual comparison by which the simulation is adjusted to the experiment.

Several recent computational studies summarized experimental Raman intensities of small molecules and compared them to *ab initio* and DFT calculations [65–67]. B3LYP was shown to reproduce absolute intensities of methane and ethane (the two most interesting test cases for this work) commonly within 20 % when combined with basis sets optimized to reproduce polarizabilities accurately [65, 66]. Pople basis sets approach the quality of these specialized basis sets when diffuse (+) and polarization functions (*) are added [67, 68], because the description of the more

Table 4.3 Gauche to all-trans abundance in percent from Eqs. 4.5 and 4.8, calculated for n-pentane and n-hexane conformers at temperatures determined from jet spectra (Figs. 4.7 and 4.8)

Conformer	Symmetry[a]	$q_{g,rot}/q_{t,rot}$	$q_{g,vib}/q_{t,vib}$	ΔE^{el}_{ref}	ΔE^{el}	ΔE^{0}	Equation 4.5 (%)	Equation 4.8 (%)
n-pentane ($T = 170$ K)								
tg	C_1	2.08	1.04	2.57	2.47	2.86	57	53
gg	C_2	1.01	0.99	4.02	3.98	5.07	5.4	5.6
g^+x^-	C_1	2.06	0.97	11.77	11.47	12.24	0.07	0.07
n-hexane ($T = 150$ K)								
ttg	C_1	2.23	0.94	2.49	2.34	2.94	40	38
tgt	C_2	1.05	1.02	2.53	2.43	2.88	22	20
tgg	C_1	2.16	0.86	3.91	3.78	5.10	6.5	6.7
gtg	C_2	1.08	1.00	4.93	4.68	5.82	2.1	1.9
g^+tg^-	C_i	2.07	1.14	5.45	5.20	6.08	1.8	1.5
ggg	C_2	0.99	0.71	5.23	5.18	7.19	0.5	0.6
tg^+x^-	C_1	2.20	0.82	11.46	11.08	11.92	0.03	0.03

ΔE^{el} denotes the electronic energy and ΔE^{0} the zero-point corrected energy relative to all-trans from B3LYP-D3/6-311++G** calculations (Turbomole v6.4 [44]). Relative electronic energies from Gruzman et al. [18] are included as reference (coupled cluster extrapolation, ΔE^{el}_{ref}). Energies are given in kJ mol^{-1}

[a] $g = 1, \sigma = 2\ (C_{2v},\ C_{2h})$
$g = 2, \sigma = 2\ (C_2)$
$g = 1, \sigma = 1\ (C_i)$
$g = 2, \sigma = 1\ (C_1)$

Table 4.4 B3LYP-D3/6-311++G** single gauche energies relative to all-trans (ΔE^0) in kJ mol^{-1} and associated gauche/trans abundance ratio $\left(\frac{N_g}{N_t}\right)$ at the temperature used in the B3LYP simulations of low-frequency spectra (Sect. 4.4), calculated using Eq. 4.8 with $\sigma_t = 2$, $\sigma_g = 1$, $g_t = 1$, and $g_g = 2$ if not stated otherwise

$n = 13$	ΔE^0	$\frac{N_g}{N_t}$ (110 K) (%)	$n = 14$	ΔE^0	$\frac{N_g}{N_t}$ (100 K) (%)	$n = 15$	ΔE^0	$\frac{N_g}{N_t}$ (110 K) (%)
Gauche 1	2.94	16	Gauche 1	2.40	22	Gauche 1	2.61	23
Gauche 2	2.74	20	Gauche 2	2.23	27	Gauche 2	2.42	28
Gauche 3	2.64	22	Gauche 3	2.13	31	Gauche 3	2.32	32
Gauche 4	2.72	21	Gauche 4	2.27	26	Gauche 4	2.47	27
Gauche 5	2.62	23	Gauche 5	2.26	26	Gauche 5	2.35	31
			Gauche 6[a]	2.28	13	Gauche 6	2.58	24
$n = 16$	ΔE^0	$\frac{N_g}{N_t}$ (90 K) (%)	$n = 17$	ΔE^0	$\frac{N_g}{N_t}$ (90 K) (%)	$n = 18$	ΔE^0	$\frac{N_g}{N_t}$ (80 K) (%)
Gauche 1	2.60	12	Gauche 1	2.63	12	Gauche 1	2.56	8
Gauche 2	2.36	17	Gauche 2	2.36	17	Gauche 2	2.23	14
Gauche 3	2.30	18	Gauche 3	2.34	18	Gauche 3	2.23	14
Gauche 4	2.40	16	Gauche 4	2.42	16	Gauche 4	2.25	14
Gauche 5	2.33	18	Gauche 5	2.30	19	Gauche 5	2.16	16
Gauche 6	2.57	13	Gauche 6	2.63	12	Gauche 6	2.50	9
Gauche 7[a]	2.70	5	Gauche 7	2.58	13	Gauche 7	2.42	11
						Gauche 8[a]	2.47	5

(continued)

Table 4.4 (continued)

$n = 19$	ΔE^0	$\frac{N_g}{N_t}$ (90 K) (%)	$n = 20$	ΔE^0	$\frac{N_g}{N_t}$ (80 K) (%)	$n = 21$	ΔE^0	$\frac{N_g}{N_t}$ (90 K) (%)
Gauche 1	2.70	11	Gauche 1	2.54	13	Gauche 1	2.74	10
Gauche 2	2.39	16	Gauche 2	2.18	22	Gauche 2	2.37	17
Gauche 3	2.44	15	Gauche 3	2.25	20	Gauche 3	2.50	14
Gauche 4	2.41	16	Gauche 4	2.20	21	Gauche 4	2.40	16
Gauche 5	2.31	18	Gauche 5	2.15	23	Gauche 5	2.36	17
Gauche 6	2.67	11	Gauche 6	2.42	16	Gauche 6	2.63	12
Gauche 7	2.46	15	Gauche 7	2.23	20	Gauche 7	2.39	16
Gauche 8	2.79	10	Gauche 8	2.63	12	Gauche 8	2.85	9
			Gauche 9[a]	2.79	5	Gauche 9	2.77	10

[a] C_2-symmetry, $\sigma_g = 2$

polarizable outer region of the molecule is better in this case. This is why the triple zeta basis set used here was augmented with diffuse and polarization function on both, heavy and light atoms (indicated by ++ and **, respectively). Beside the fair prediction of absolute intensities, it should be noted that the overall polarizability of alkanes is well described by additive bond polarizabilities [69, 70], which suggests that *relative* intensities of different conformers predicted by the B3LYP approach might involve an uncertainty of less then 20 %. The strong predictive power of B3LYP regarding relative Raman intensities was recently demonstrated by Zvereva et al. [67].

The uncertainty due to the poorly defined vibrational temperature is critical merely at low-frequencies. It may be assessed by the following sensitivity analysis: For vibrations at 100 and 200 cm^{-1},[9] varying the temperature from 100 K to lower and upper bounds of 50 and 200 K, respectively, changes the scattering cross-section ratios by at most 20 %. This can be calculated from the temperature dependence of the scattering cross-section, or specifically the ratio:

$$\frac{\sigma_k'}{\sigma_l'} \propto \left[1 - \exp\left(-\frac{hc\tilde{\nu}_l}{kT}\right)\right] / \left[1 - \exp\left(-\frac{hc\tilde{\nu}_k}{kT}\right)\right],$$

for two vibrations with wavenumbers $\tilde{\nu}_{k,l}$ (valid in the harmonic approximation, see Sect. 2.1).

The uncertainty due to the visual matching of simulation and experiment is probably in the same range. The overall uncertainty in the conformer fractions due to these independent inaccuracies is estimated to be about 30 %.

The next paragraphs will discuss the quality of the simulations using concrete examples of the small alkanes pentane and hexane, as well as tridecane, the shortest alkane investigated in the context of alkane-folding.

4.3.5 Case Study: Pentane and Hexane in the Low-Frequency Region

Experimental Raman spectra of pentane and hexane are compared to simulated spectra in Figs. 4.7 and 4.8. The bottom graphs show unweighted stem plots of relative scattering cross-sections. Note that stem plots show vibrations only if they have at least 0.5 % of the accordion vibration intensity,[10] for better clarity. Gas phase spectra and the hexane jet-spectrum were measured by Tobias Wassermann [6]. For gas phase spectra, measured by filling the vacuum chamber of the curry-jet with pentane or hexane mixed with helium, the temperature is known (≈300 K), but the more pronounced rotational structure complicates the comparison to the simulation, in

[9] This is close to important all-trans accordion vibrations and hairpin vibrations, see later in the text.

[10] Pentane: 1.77×10^{-34} $m^2\,sr^{-1}$ (at 396 cm^{-1}); hexane: 1.74×10^{-34} $m^2\,sr^{-1}$ (at 366 cm^{-1}).

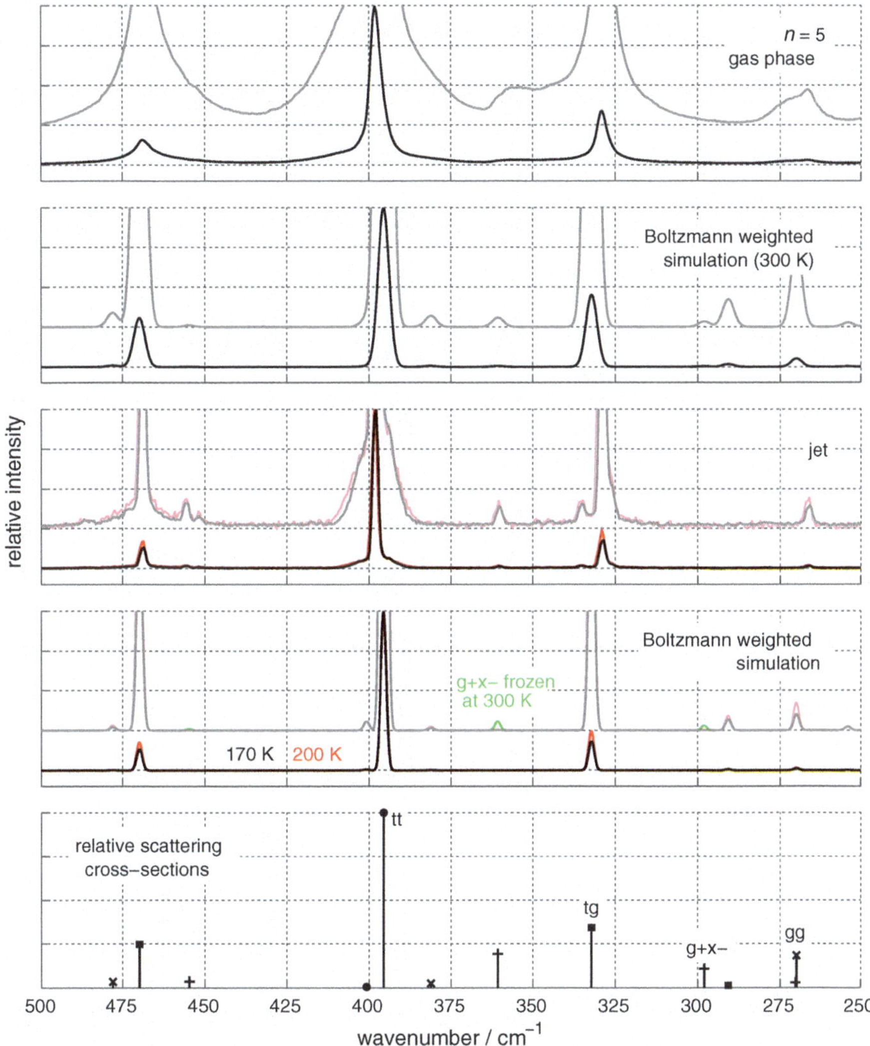

Fig. 4.7 Simulated *n*-pentane spectra, set against Raman gas phase [6] and jet-spectra. Magnified spectra (10×) are drawn in *light colors*. Jet-spectra drawn in *black* and *gray*: nozzle at room temperature; jet spectrum drawn in *red*: nozzle at 130 °C

particular when bands are close to each other and the rotational branches overlap. In case of jet-spectra, the temperature is a free parameter of the simulation. The simulated spectrum is scaled to match the experimental all-trans accordion vibration band integral, and the conformational temperature is varied until a visual comparison between experiment and simulation shows satisfactory conformity. The error of the so estimated conformational temperature, which is indicated in the figures, is kept

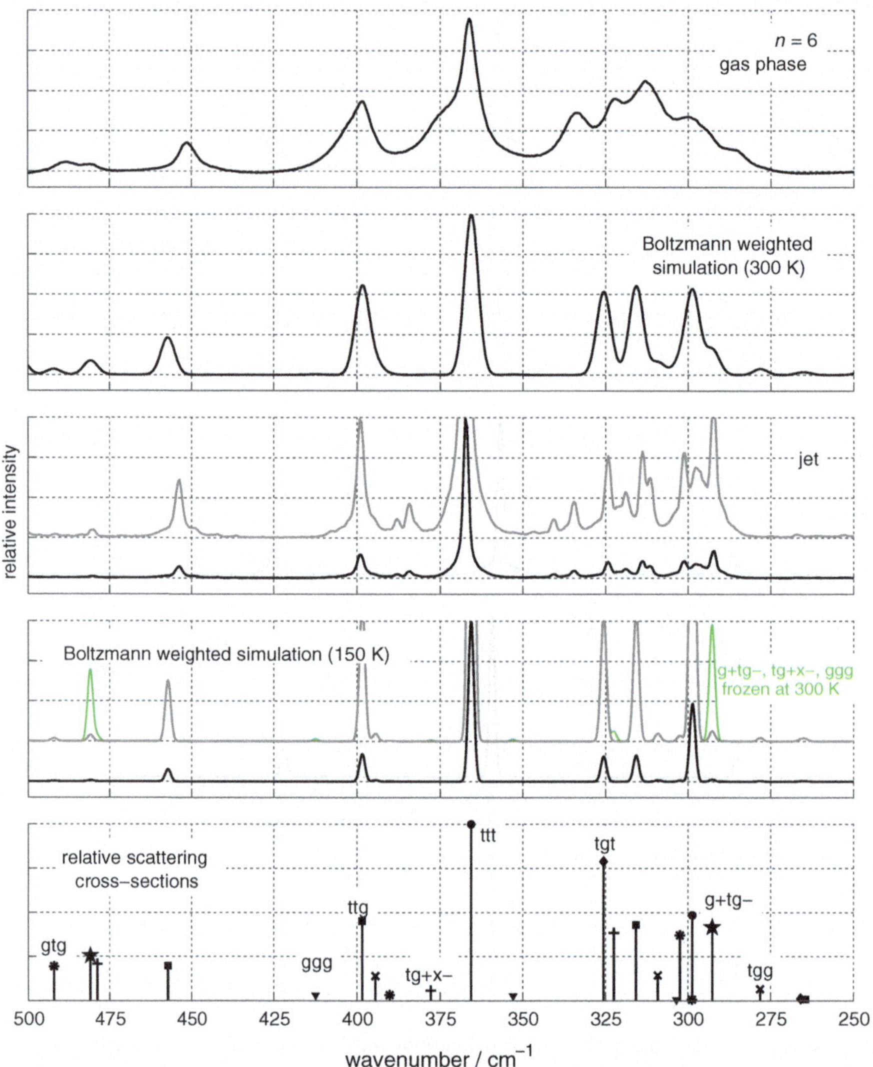

Fig. 4.8 Simulated n-hexane spectra, set against Raman gas phase and jetspectra [6]. Magnified spectra (5×) are drawn in *gray*

relatively low by the exponential temperature dependence (about ±15 K), but in particular the error due to approximations (largely neglected differences in partition functions) is harder to assess.

Pentane

The pentane simulation is "complete" in the sense that all stable isomers are considered. In the experimental gas phase spectrum, all isomers can be identified, but not all bands of the simulated spectrum are observed due to overlap with rotational branches. Jet-spectra compare more favorably. Simulated and measured single gauche (tg) and all-trans (tt) bands are virtually a perfect match, except for the weak single gauche band at 291 cm^{-1} which overshoots in the simulated spectra. The small band from the double gauche (gg) conformer at $\approx$270 cm^{-1} is also well reproduced, which indicates that gg and gt conformers share a uniform freezing temperature. Overall, the simulation performs very well in this case.

Some bands from the experimental jet spectrum are missing in the simulated spectrum. In general, such bands can stem from anharmonic resonances which are not described in the harmonic approximation, hot bands which are not shifted from ground state transitions in the harmonic approximation, or aggregates which are simply not included in the simulation. In this particular case, at least the weak signal at $\approx$360 cm^{-1} might have an origin which is actually covered by the simulation, but does not contribute due to the assumption of thermal equilibrium. It might belong to the high energy *syn*-pentane conformer (g^+x^-). If so, this conformer freezes earlier than the single and double gauche conformer, because the conformational temperature of the latter (<200 K) would imply a much lower g^+x^- abundance than observed in the experimental jet-spectrum. Freezing of the g^+x^- conformer at 300 K is considered in the green simulated spectrum. The experimental band at 455 cm^{-1} matches the wavenumber of another g^+x^- vibration, but its intensity is too high for such an assignment, even when freezing is considered.

Aggregate bands can be excluded based on the findings from Ref. [6]. There, only one band at 480 cm^{-1} showed the characteristic cluster behavior with varying measurement conditions. This band is missing in spectra from this work, which is beneficial considering that cluster formation is not intended. This might be explained by lower substance concentration and stagnation pressure used here.

Hexane

In case of hexane, the missing rotational structure in the simulated gas phase spectrum is a more obvious disadvantage, especially in the region around 300 cm^{-1} where several bands overlap. The simulation of jet-spectra with simple Gaussian curves works better, but the number of bands missing in the simulation is now much higher. One obvious reason is that not all stable isomers are considered in the simulation in this case. Excluded conformers are of high energy, but early freezing (like in case of the pentane g^+x^- conformer) might lead to significant contributions to the Raman jet-spectrum. Letting the considered higher energy conformers freeze at 300 K (green spectrum) leads to a better agreement in case of some bands, but worse agreement in case of others (g+tg−), such that no consistent prediction is possible. Also, the simulations suffers from an overestimated mode coupling in case of the all-trans (ttt) conformer,[11] such that the all-trans vibration at 300 cm^{-1} is too intense.

Clearly, the more complicated hexane spectrum is less well reproduced by the harmonic simulation than the pentane spectrum, but this trend does not extend to longer alkanes with chain lengths of at least 13 carbon atoms, studied in the context of chain folding. Those can occupy a much higher number of stable multi gauche

[11] A general discussion of this aspect will be given in more detail in Sect. 6.1.

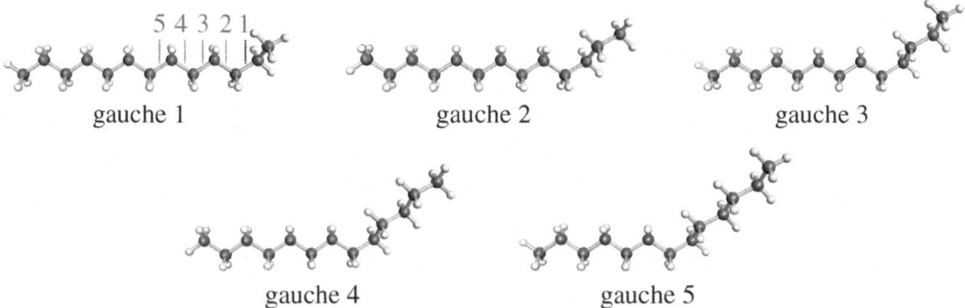

Fig. 4.9 Single gauche nomenclature shown for the example of tridecane

conformations, causing the corresponding multi gauche bands to overlap to broad and unstructured features. Single gauche bands, on the other hand, stand out much clearer. The last example, tridecane, will show that beside the all-trans conformer only the single gauche conformers shown in Fig. 4.9 need to be considered to closely reproduce most of the features of the corresponding experimental jet-spectrum.

4.3.6 Case study: Tridecane in the Low-Frequency Region

In Fig. 4.10, a B3LYP-D3/6-311++G** simulation of the low-frequency region of tridecane ($n = 13$) is compared to experimental Raman jet spectra. When examining Table 4.1 (p. 41), one finds $n-3 = 10$ single gauche conformers for tridecane, one for each possible position of the gauche conformation along the chain. However, these ten conformers include pairs of enantiomers, and only five single gauche conformers show a unique vibrational Raman spectrum. For tridecane, the five single gauche conformers which need to be considered have gauche torsional angles in position 1–5, when the carbon-carbon bonds open to torsional isomerism are numbered with an index running from 1 to 10. Conformers with the gauche conformation in position 6–10 are enantiomers of conformers with the gauche conformation in position 5–1, respectively, which are illustrated in Fig. 4.9. In Fig. 4.10, unweighted stem plots are included to help identify the specific conformers.

The conformational temperature used to weight single gauche conformers in the simulated spectrum shown in Fig. 4.10 is 110 K, which is seen to be suited for the weighting of all different single gauche conformers. This is also true for longer alkanes, shown later in the text, and underlines an equilibrated or uniformly frozen single gauche/all-trans distribution as expected. The hairpin conformer is weighted manually to 10 % of the all-trans abundance and does not contribute to the experimental spectrum according to this comparison with the B3LYP prediction.

The rich structure of the jet spectrum, which will be assigned in Sect. 4.4, provides a solid ground to assess the quality of the B3LYP predictions in the important size range $n = 13$–21. Without examining the agreement of experimental and predicted

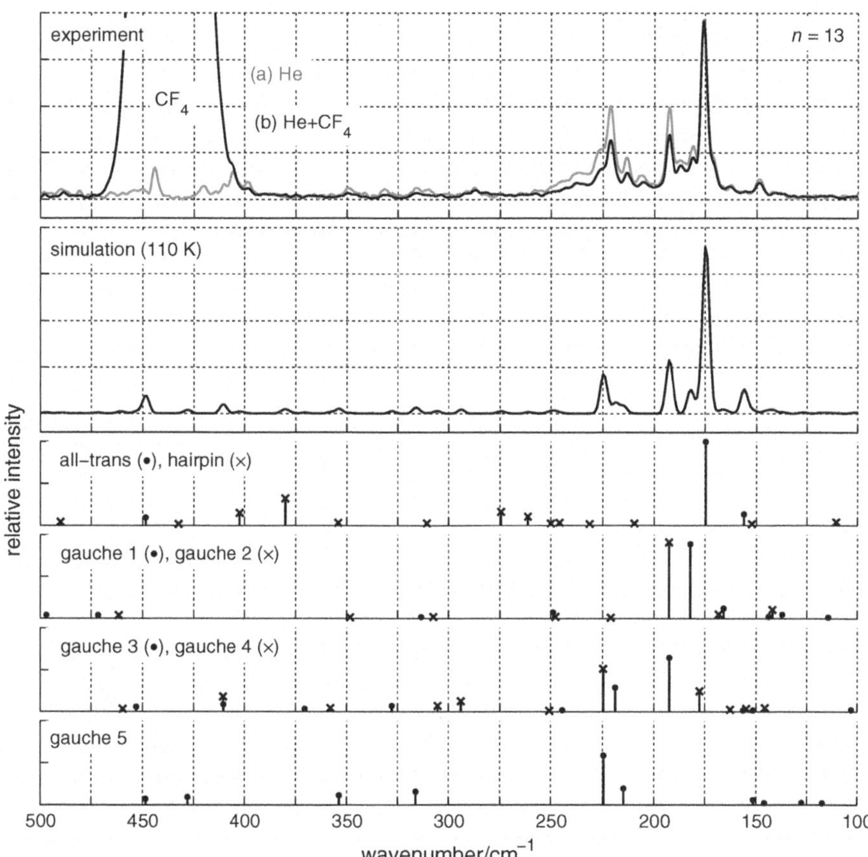

Fig. 4.10 Comparison of averaged jet-cooled Raman spectra of tridecane in the low-frequency region to a B3LYP-D3/6-311++G** simulation. Experimental spectra are Savitzky-Golay filtered (9 pt.). Stem plots correspond to unweighted scattering cross sections relative to the all-trans accordion vibration. Only vibrations with at least 1 % of the accordion vibration intensity are included. *Measurement conditions* **a** expansion in He, $\vartheta_s = 37\,°C$, $p_0 = 0.5\,bar$, $p_b = 0.9\,mbar$, $d_n = 1\,mm$, exposure 24× 300 s, **b** expansion in He + 4 % CF_4, $\vartheta_s = 37\,°C$, $p_0 = 0.9\,bar$, $p_b = 1.3\,mbar$, $d_n = 1\,mm$, exposure 24× 300 s. *Calculation* effective temperature used in single gauche to all-trans weighting = 110 K, hairpin to all-trans weighting = 0.1:1, wavenumber scaling factor = 0.99. Reference II—Reproduced by permission of The Royal Society of Chemistry

wavenumbers quantitatively, it is obvious that the simulated spectrum matches the experimental spectrum closely. This holds true for longer alkanes with a few exceptions. The spectra of C–C stretching vibrations are described similarly well, but the simulation fails in the region of C–H stretching vibrations, which is not surprising because of the large impact of Fermi interaction which occurs in this wavenumber range in alkyl systems [15, 71–73].

4.4 Experimental Raman Jet Spectra

This section will provide a detailed assignment of Raman bands observed for alkanes isolated in supersonic jet expansions. The assignment is divided into three spectral regions, characteristic for skeletal vibrations (low-frequency region, from the Rayleigh line up to $\approx$600 cm^{-1}), C–C stretching and C–H deformation vibrations (800–1400 cm^{-1}), and C–H stretching vibrations (2800–3000 cm^{-1}).

4.4.1 Low-Frequency Region

The low-frequency region jet spectra are analyzed aided by quantum chemical simulations for the most part. Comparison with the literature (liquid and solid state spectra) is less helpful, since the available spectra correspond either to alkanes in a much more disordered state (the liquid state at or above room temperature) or alkanes almost entirely ordered in a crystal, where the hydrocarbon chains occupy the all-trans conformation. Raman jet spectra are largely characterized by the strong all-trans accordion vibration, but show also distinct bands from single gauche conformers, which puts their appearance in between solid and liquid phase spectra. Besides the all-trans LAM-1 or accordion vibration and accordion-like vibrations from single gauche conformers, the all-trans LAM-3 and associated single gauche vibrations are features commonly found. In general, spectral features of the low-frequency region are found to be most sensitive to both, the chain length and conformation, which is why every alkane is discussed separately before the spectra are set against each other and examined for chain length dependent trends. Because of spectral similarities, the alkanes from $n = 13$–21 will be discussed in groups of three.

Tridecane, Tetradecane, and Pentadecane

The normal alkanes up to pentadecane were already measured in the spectral region close to the accordion vibration with a satisfactory signal-to-noise ratio using the curry-jet setup at room temperature [6, 8]. No spectral markers of hairpin conformers could be identified in these studies. In this work, a higher signal-to-noise ratio was achieved at 1mm nozzle distance due to the usage of the heatable nozzle, but no definite hairpin contribution to the Raman spectra was found either. Despite the missing hairpin evidence, an assignment and comparison to simulated spectra is interesting to elucidate the spectral development with the chain length, and to probe the quality of the quantum chemical predictions. By comparison to the simulated spectra, it can be shown that for $n = 13$–15, vibrational bands aside from the accordion vibration band arise almost entirely due to single gauche conformers. For tridecane, this was indicated in Sect. 4.3.6, where most of the spectral features were seen to be covered by the all-trans/single gauche simulation. However, a thorough assignment of the bands was not provided and is given in the following.

Figure 4.10 shows low-frequency spectra of tridecane expanded in different carrier gas mixtures and from different stagnation pressures (p_0) but otherwise matching experimental conditions. The He/CF_4 expansion reaches a lower effective temperature letting all but the all-trans bands decrease when the spectra are scaled to the accordion vibration (176 cm^{-1}). Such relaxation experiments allow to identify all-trans contributions, but to tell single gauche bands apart from each other, the simulated spectrum needs to be considered. Double and higher gauche conformers with higher energies can be realized in numerous ways and should contribute to a broad background rather than to sharp bands. The tridecane jet spectra show a broad background in the region 200–250 cm^{-1}, which is assigned to multi gauche conformers accordingly. Raman active skeletal vibrations of disordered alkanes are known to accumulate in this wavenumber region if $n > 10$ [40]. The relaxation experiment helps to identify the weak band at 149 cm^{-1} as an all-trans vibration. The B3LYP calculation shows that it is a totally-symmetric transversal acoustical mode (TAM), belonging to the same C–C–C bending frequency branch as the LAM-1. When such a vibration of appropriate symmetry lies close in energy to the LAM-1, both vibrations mix, yielding vibrations with partial LAM/TAM character and redistributed intensity—a point which is of further importance in Chap. 6 and will be discussed there in more detail. The mode mixing follows naturally from solving the systems' equations of motion in the harmonic approximation, and is thus predicted by the harmonic frequency calculation. Comparing theory and experiment in this aspect shows that the mixing is predicted to be more pronounced than actually observed, also indicated by the wavenumber difference of the mixing TAM and accordion vibration being too small in the calculation. Such subtle inaccuracies can grow to more drastic deviations, as seen later in case of the low-frequency eicosane spectrum. As the last all-trans contribution in the low-frequency region, the simulated spectrum helps to identify the LAM-3 at 444 cm^{-1}, not observable in the He + 4 % CF_4 expansion due to overlap with a CF_4 deformation vibration.

Low-frequency Raman spectra of single gauche conformers are more crowded. Twisting one of the C–C bonds of the all-trans conformer into a gauche conformation breaks the symmetry of the system, so that mode mixing will not be restricted by symmetry anymore, and the accordion vibration intensity will scatter over several vibrations. In essence, this is what is found from normal mode calculations and observed in the experiment. Studying the calculated normal modes of single gauche skeletal vibrations, some qualitative rules can be found: The intensity of single gauche vibrations close to the accordion vibration depends on how much character of the accordion vibration they contain. Bands with high Raman intensity correspond to vibrations with a high amount of longitudinal stretching of the whole molecule, such bands with weak Raman intensity have more character of transversal acoustic modes. Understandably, a gauche bond close to the chain end changes the vibrational spectrum compared to all-trans relatively little, while a gauche bond close to the chain center has a much larger effect. This is seen in Fig. 4.10 when the two limits—a gauche bond in position 1 and a gauche bond in position 5—are compared. The former conformer exhibits a spectrum closely related to the all-trans spectrum, with one dominating band shifted slightly from the accordion vibration band, while the latter conformer involves vibrations shifted to considerably higher wavenumber and more fragmented intensity. This trend is in accordance with the remaining

Table 4.5 Assignment of experimental low-frequency Raman bands: tridecane, tetradecane, and pentadecane

Tridecane		Tetradecane		Pentadecane	
Wavenumber	Assignment	Wavenumber	Assignment	Wavenumber	Assignment
149 (w)	All-trans[a]	158 (m)	All-trans[b]	154 (s)	All-trans[c]
176 (s)	All-trans[c]	165 (s)	All-trans[b]	162–179 (w,b)	Gauche 1–3
181 (w)	Gauche 1	176 (w)	Gauche 2, *1, 5*	202 (m)	Gauche 4, 5
193 (m)	Gauche 2, 3	185 (vw)	Gauche 3	210 (vw)	*Gauche 3*
205 (vw)	*Gauche 5*	209 (m)	Gauche 4, 6	217 (w)	*Gauche 6*
213 (w)	Gauche 3	216 (m)	Gauche 5, *3*	398 (m)	All-trans[d], Gauche 2–6
222 (m)	Gauche 4, 5	393 (vw)	*Gauche 5*		
406 (w)	*Gauche 3, 4*	412 (m)	Gauche 3, 4	433 (vw)	*Gauche 1*
420 (w)	*Gauche 5*	433 (m)	All-trans[d]		
444 (w)	All-trans[d]				

Tentative assignments are printed in italics. Wavenumbers are given in cm^{-1}, relative intensities in parentheses (b = broad, v = very, w = weak, m = medium, s = strong). The number characterizing gauche conformers indicates the position of the gauche bond (see Fig. 4.9)

[a] TAM with LAM-1 character

[b] TAM / LAM-1 Fermi resonance pair

[c] LAM-1

[d] LAM-3

single gauche conformers. Additionally, in contrast to the all-trans conformer, single gauche conformers contribute to the spectral region of the LAM-2 between 250 and 400 cm^{-1}, because symmetry constraints are lifted. This gives rise to the very weak bands observed in the jet spectrum in this region. Single gauche bands scattering around 400–450 cm^{-1} are associated with the all-trans LAM-3 and share a similar displacement along the normal coordinate. Several single gauche vibrations may become Raman active, analogous to vibrations associated with the LAM-1.

The assignment of Raman bands to specific conformers is given in Table 4.5. Certainly, some of the assignments of weaker Raman bands must be considered rather tentative. Interesting assignments regard the strong bands at 193 and 222 cm^{-1}. They seem to suggest a quite high single gauche abundance, but are actually the result of overlapping gauche 2/gauche 3 and gauche 4/gauche 5 bands, respectively. As stated above, the hairpin conformer is not assigned yet. In principle, the weak band at 399 cm^{-1} is compatible with a hairpin vibration, but another (more intense) hairpin band, predicted to show up at somewhat lower wavenumber, is missing.

Raman jet spectra of tetradecane together with a B3LYP simulation are shown in Fig. 4.11. The accordion vibration (159, 166 cm^{-1}) is found to be perturbed by a combination tone of two TAMs in Fermi resonance [6, 8]. The affiliation of the 159 cm^{-1} band to all-trans is verified by comparison of He expansions and He/CF_4 expansions at lower effective temperatures. This anharmonic coupling is not covered by the harmonically approximated simulation, which shows a single strong all-trans accordion vibration band. The single gauche pattern is correctly captured by the simulation, but the bands at the high frequency flank of the accordion vibration are

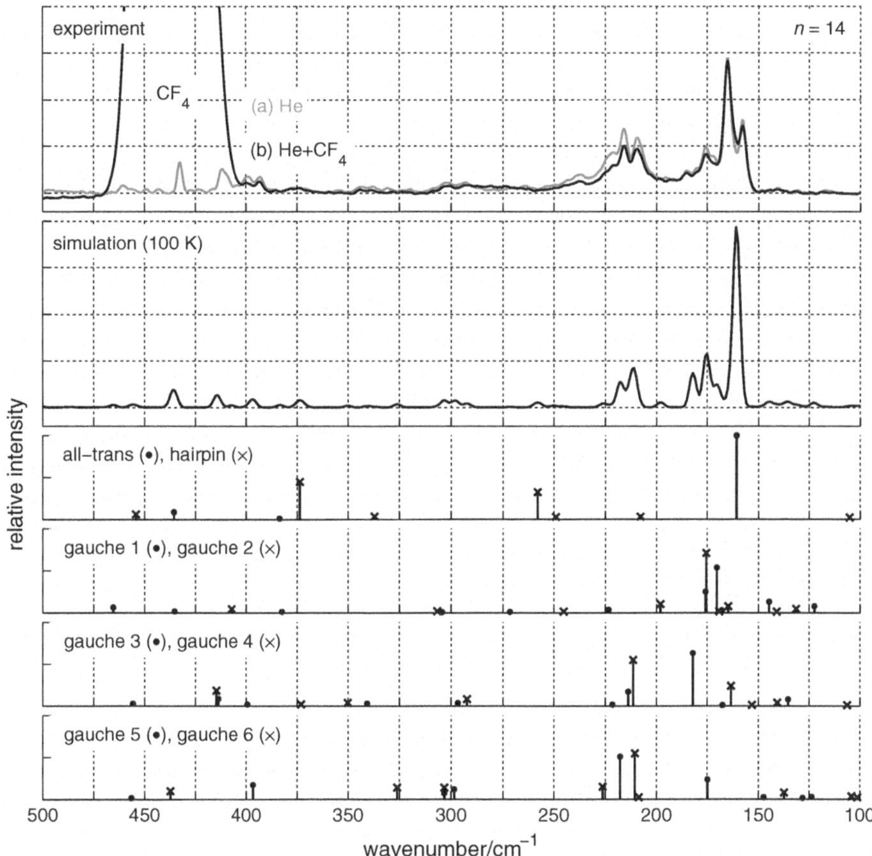

Fig. 4.11 Comparison of averaged jet-cooled Raman spectra of tetradecane in the low-frequency region to a B3LYP-D3/6-311++G** simulation. Experimental spectra are Savitzky-Golay filtered (9 pt.). *Measurement conditions* **a** expansion in He, $\vartheta_s = 45\ °C$, $p_0 = 0.5$ bar, $p_b = 0.8$ mbar, $d_n = 1$ mm, exposure 24× 300 s, (**b**) expansion in He + 4 % CF_4, $\vartheta_s = 45\ °C$, $p_0 = 0.85$ bar, $p_b = 1.2$ mbar, $d_n = 1$ mm, exposure 12 × 300 s. *Calculation* effective temperature used in single gauche to all-trans weighting = 100 K, hairpin to all-trans weighting = 0.1:1, wavenumber scaling factor = 0.99

much more blurred into a broad background, perhaps because the Fermi resonance found for all-trans partly remains in single gauche conformers with the gauche bond close to the chain end, leading to a more congested spectrum. The hairpin conformer should contribute to the Raman spectrum mainly at ≈375 and ≈255 cm^{-1}, according to the simulation. At the lower wavenumber, no signals are found, but at ≈375 cm^{-1}, the experimental spectra show a very weak and broad elevation, reproducible from the He to the He/CF_4 expansion, which could indicate the hairpin conformer. However, the signal is too weak to draw convincing conclusions. If the hairpin conformer is present, its abundance should be lower than the 10 % relative to all-trans set in the simulation.

A shoulder in between the accordion vibration Fermi resonance pair ($\approx$163 cm^{-1}) was found to vary notably with the expansion conditions. Judging from the wavenumber, it is compatible with a gauche 4 vibration, but the relaxation behavior upon CF_4 addition rather suggests an all-trans assignment. The shoulder is sensitive to the concentration of tetradecane and is a rare case of a low-frequency feature compatible with a cluster assignment (stronger intensity at higher concentration).

Low-frequency pentadecane spectra set against B3LYP simulations are shown in Fig. 4.12. The shortcomings of neglecting multi gauche conformers become more obvious at this chain length. The accordion vibration band (154 cm^{-1}) and single gauche bands around 200 cm^{-1} sit on broader multi gauche bands, which are necessarily missing in the simulated spectra. The characteristic bands aside from the accordion vibration are overlapping single gauche bands at 202 cm^{-1} and the LAM-3 at 398 cm^{-1}. The LAM-3 band is unusually strong because it overlaps with bands from the conformers gauche 2–6. The structured band at the high-frequency flank of the accordion vibration band (162–179 cm^{-1}) is not resolved, but the main contributors can be identified as conformers gauche 1, 2, and 3. According to the B3LYP prediction, hairpin vibrations should be scattered mainly around 250 cm^{-1}. A relatively isolated hairpin band is predicted to show up at 366 cm^{-1}, but the experiment shows no clear indication of corresponding bands.

Hexadecane, Heptadecane, and Octadecane

For hexadecane and the longer following alkanes, single gauche bands are less prominent, and the broad background around 200 cm^{-1}, characteristic for multi gauche conformers, gains intensity. To aid the visual comparison of experiment and simulation, this spectral feature is accounted for with one additional broad Gaussian curve, added to the simulated spectrum. Its height, width, and position are taken from a fit of the He expansion spectrum.

Hexadecane Raman jet spectra are shown in Fig. 4.13. The dominating all-trans accordion vibration band is found at 143 cm^{-1}. Beside the all-trans LAM-3 band (389 cm^{-1}), two eye-catching bands of comparable intensity are found at 162 and 194 cm^{-1}. According to the B3LYP calculations, the latter is a superposition of gauche 5 and 6 bands, which exhibit Raman-active vibrations at very similar wavenumbers. Like in the case of some shorter alkane single gauche bands, the resulting band falsely suggests a single conformer origin at the present low resolution. The band at 162 cm^{-1} is assigned to the gauche 3 conformer. Gauche 7 could intensify this band, but it has an energetic disadvantage (Table 4.4) and a statistical disadvantage (C_2-symmetry). Thus, it rather adds to the broader background at $\approx$160 cm^{-1}, together with gauche 1 and 2. The Raman intensity of gauche 3 is largely confined to one vibration and the conformational energy relative to all-trans is the lowest among the single gauche conformers, allowing the corresponding band to contrast with the more diffuse character of the other single gauche bands. This particular agreement between experiment and calculations could be accidental, but also points to a robust

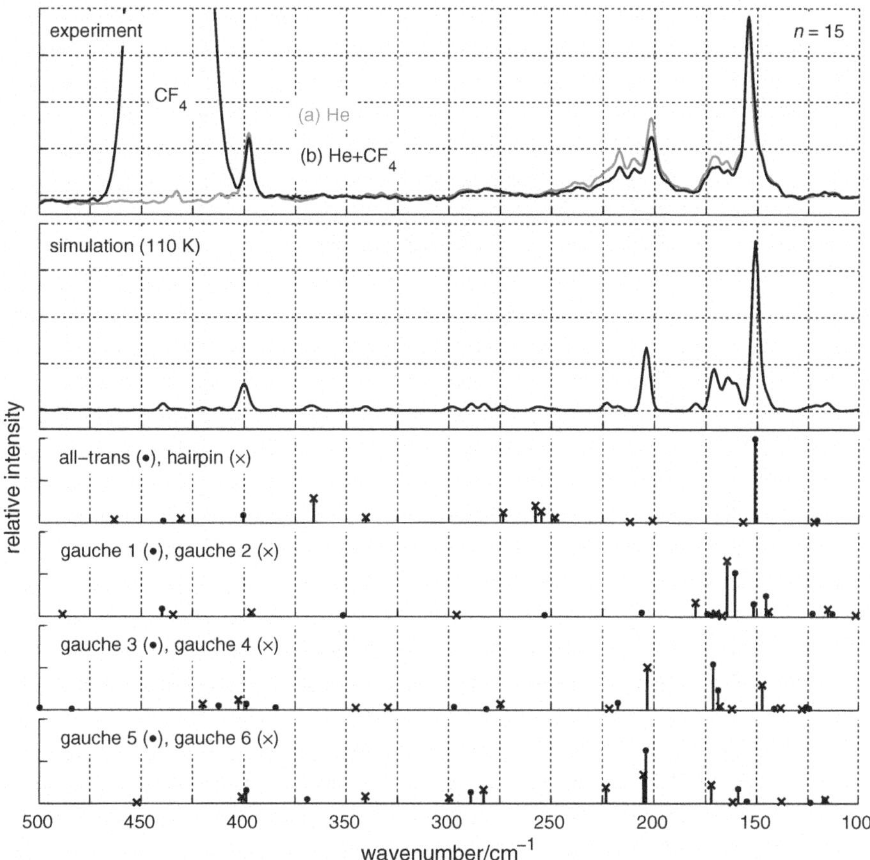

Fig. 4.12 Comparison of averaged jet-cooled Raman spectra of pentadecane in the low-frequency region to a B3LYP-D3/6-311++G** simulation. Experimental spectra are Savitzky-Golay filtered (9 pt.). *Measurement conditions* **a** expansion in He, $\vartheta_s = 55\,°C$, $p_0 = 0.5\,\text{bar}$, $p_b = 0.8$–0.9 mbar, $d_n = 1$ mm, exposure $11\times$ 300 s, **b** expansion in He + 4 % CF_4, $\vartheta_s = 55\,°C$, $p_0 = 0.85\,\text{bar}$, $p_b = 1.2$ mbar, $d_n = 1$ mm, exposure $12\times$ 300 s. *Calculation* effective temperature used in single gauche to all-trans weighting = 110 K, hairpin to all-trans weighting = 0.1:1, wavenumber scaling factor = 0.99

prediction of Raman intensities and single gauche conformational energies, provided by the applied B3LYP approach.

The most Raman intense hairpin vibration predicted at $268\,\text{cm}^{-1}$, together with some weaker vibrations close in wavenumber, seem to have a corresponding band in the experimental jet spectra. Again, the band is very weak and broad so that the hairpin abundance must be very low, if the experimental evidence stems from this conformer. Gauche 5–7 also possess some weaker vibrations between 250–275 cm^{-1} and an unspecific multi gauche assignment cannot be excluded in any case.

Table 4.6 Assignment of experimental low-frequency Raman bands: hexadecane, heptadecane, and octadecane

Hexadecane		Heptadecane		Octadecane	
Wavenumber	Assignment	Wavenumber	Assignment	Wavenumber	Assignment
143 (s)	All-trans[a]	136 (s)	All-trans[a]	127 (s)	All-trans[a]
162 (w)	Gauche 3, *1, 2, 7*	143 (w)	Gauche 1, 3	142 (w)	Gauche 2–4, *1, 5*
194 (w)	Gauche 5, 6	158 (w,b)	Gauche 3, *2, 4*	149 (vw)	*Gauche 2, 3, 8*
202 (vw)	*Gauche 4*	172 (w)	Gauche 5	156 (vw)	Gauche 5, *7*
214 (vw)	*Gauche 7*	187 (w)	Gauche 6	170 (vw)	Gauche 6
389 (m)	All-trans[b]	206 (vw)	Gauche 7	198 (vw,b)	Gauche 4, 5
		266 (vw,b)	*Gauche 4–6*	243 (w)	Hairpin
		350 (vw)	All-trans[c], *Gauche 7*	352 (m)	All-trans[b]
		360 (vw)	Gauche 5		
		378 (w,b)	All-trans[b], *Gauche 4*		
		395 (vw,b)	Gauche 3		

Tentative assignments are printed in italics. Wavenumbers are given in cm^{-1}, relative intensities in parentheses (b = broad, v = very, w = weak, m = medium, s = strong). The number characterizing gauche conformers indicates the position of the gauche bond (see Fig. 4.9)

[a] LAM-1; [b] LAM-3; [c] TAM with LAM-3 character;

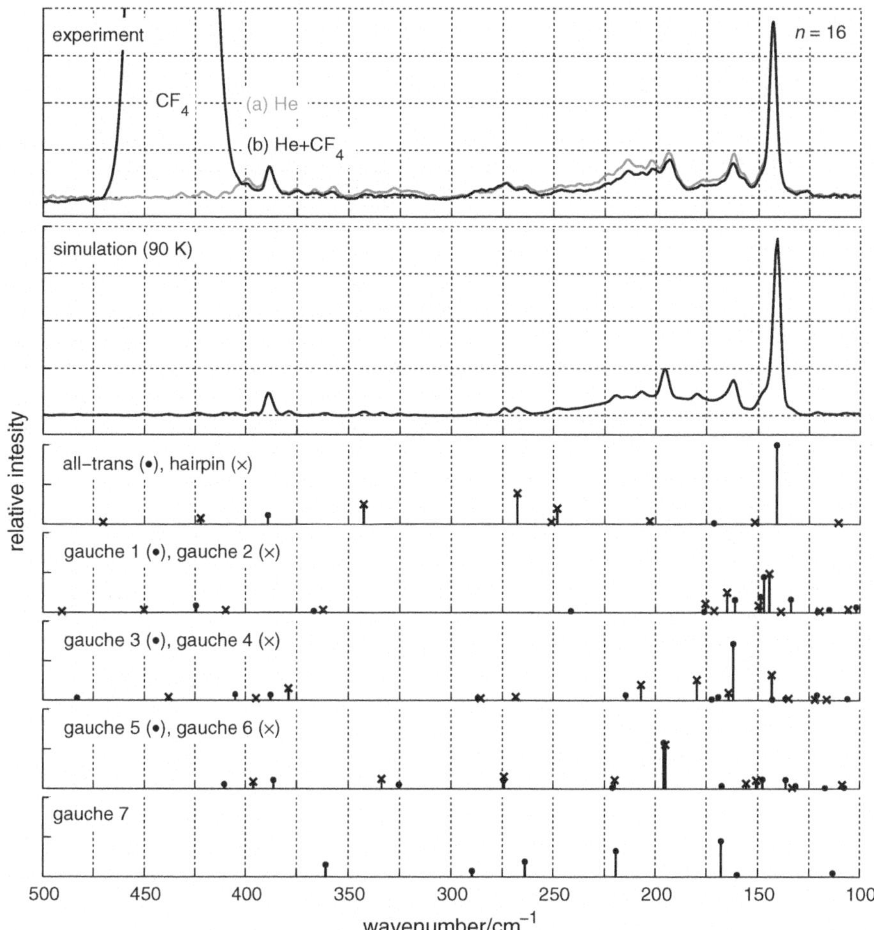

Fig. 4.13 Comparison of averaged jet-cooled Raman spectra of hexadecane in the low-frequency region to a B3LYP-D3/6-311++G** simulation. Experimental spectra are Savitzky-Golay filtered (9 pt.). *Measurement conditions* **a** expansion in He, $\vartheta_s = 65\,°\mathrm{C}$, $p_0 = 0.5\,\mathrm{bar}$, $p_b = 0.9\,\mathrm{mbar}$, $d_n = 1\,\mathrm{mm}$, exposure 18× 300 s, **b** expansion in He + 4% CF_4, $\vartheta_s = 65\,°\mathrm{C}$, $p_0 = 0.9$ bar, $p_b = 1.3$ mbar, $d_n = 1$ mm, exposure 6× 300 s. *Calculation* effective temperature used in single gauche to all-trans weighting = 90 K, hairpin to all-trans weighting = 0.1:1, wavenumber scaling factor = 0.99

Heptadecane, Fig. 4.14, exhibits low-frequency Raman jet spectra quite similar to hexadecane with a somewhat more complex single gauche band pattern but a likewise dominant all-trans accordion vibration band (136 cm^{-1}). The LAM-3 band at 378 cm^{-1} is rather faint because it mixes with an a_1 symmetric TAM at 350 cm^{-1}. The robust B3LYP prediction shows that relatively strong single gauche bands are confined to wavenumbers close to the accordion vibration band up to $\approx$210 cm^{-1}. Raman bands from the hairpin conformer are predicted to be centered

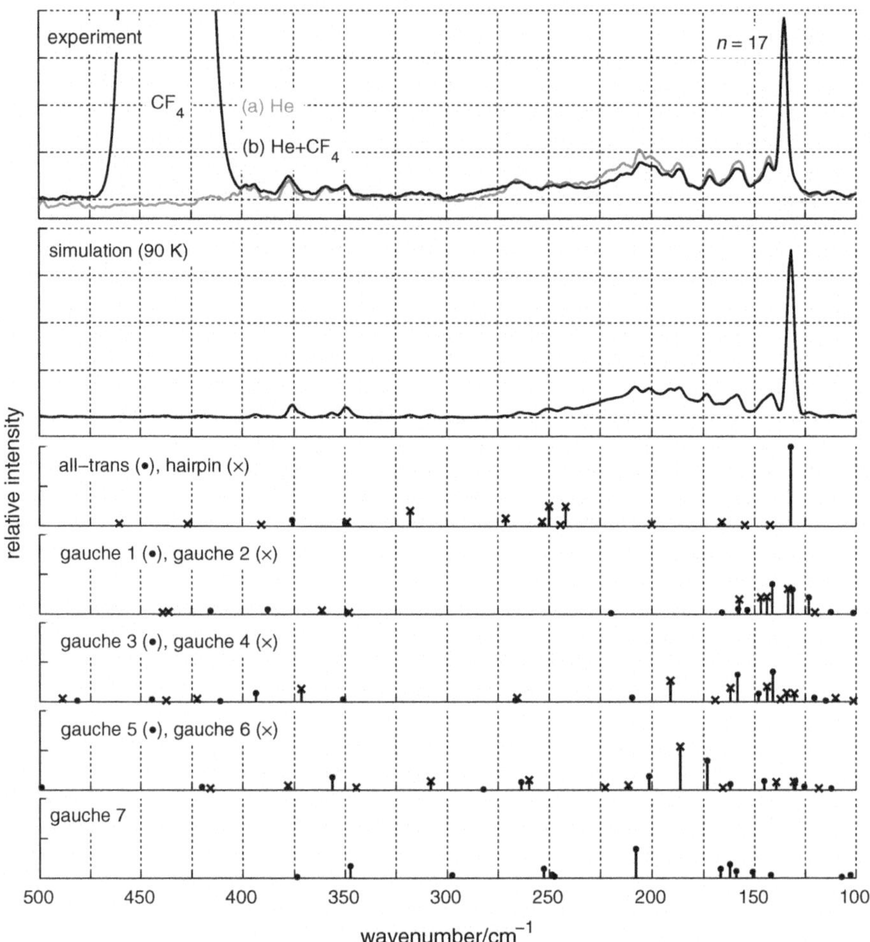

Fig. 4.14 Comparison of averaged jet-cooled Raman spectra of heptadecane in the low-frequency region to a B3LYP-D3/6-311++G** simulation. Experimental spectra are Savitzky-Golay filtered (9 pt.). *Measurement conditions* **a** expansion in He, $\vartheta_s = 75\,°C$, $p_0 = 0.5$ bar, $p_b = 0.9$ mbar, $d_n = 1$ mm, exposure 6× 300 s, **b** expansion in He + 4 % CF_4, $\vartheta_s = 75\,°C$, $p_0 = 0.9$ bar, $p_b = 1.3$ mbar, $d_n = 1$ mm, exposure 6× 300 s. *Calculation* effective temperature used in single gauche to all-trans weighting = 90 K, hairpin to all-trans weighting = 0.1:1, wavenumber scaling factor = 0.98

around 250 cm^{-1} and might have very weak and broad experimental counterparts, but as for hexadecane, the evidence does not allow a definite hairpin assignment. If the weak spectral features stem from the hairpin conformer, its abundance must still be very low.

Octadecane Raman jet spectra are set against a simulation in Fig. 4.15. Prominent all-trans bands at 127 and 352 cm^{-1} (LAM-1 and 3, respectively) contrast with weak single gauche bands between ≈125–200 cm^{-1}. Except for gauche 6, each single

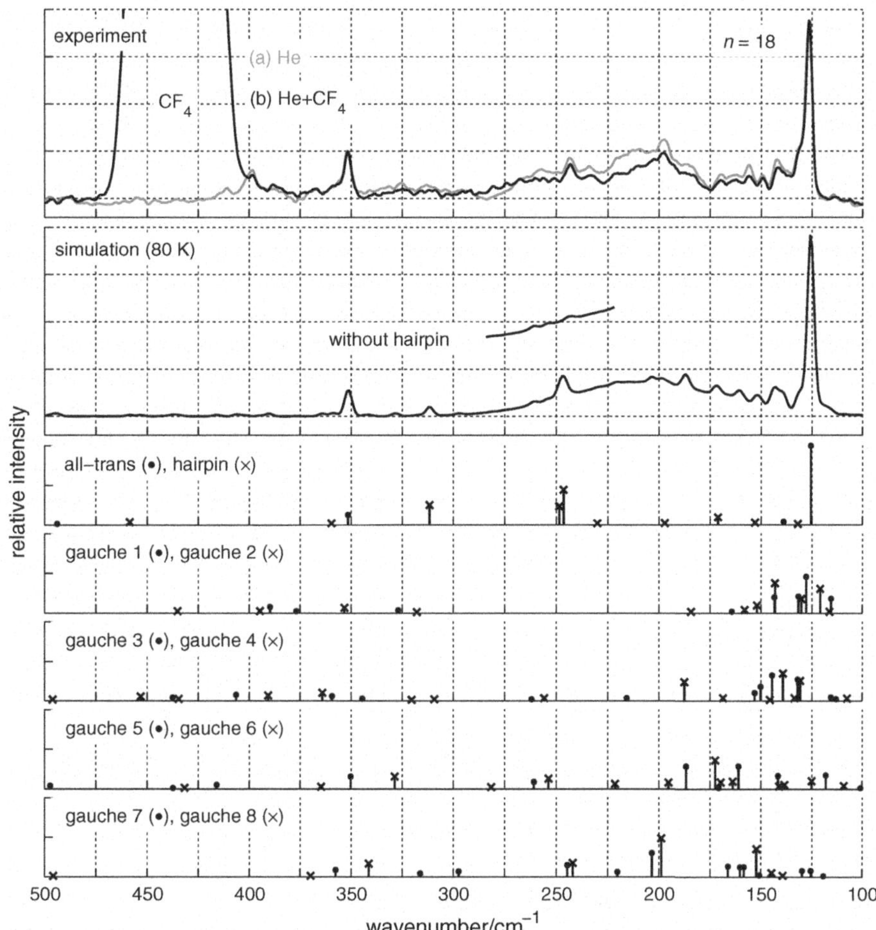

Fig. 4.15 Comparison of averaged jet-cooled Raman spectra of octadecane in the low-frequency region to a B3LYP-D3/6-311++G** simulation. Experimental spectra are Savitzky-Golay filtered (9 pt.). *Measurement conditions* **a** expansion in He, $\vartheta_s = 85\text{–}97\,°\text{C}$, $p_0 = 0.5\,\text{bar}$, $p_b = 0.9$ mbar, $d_n = 1$ mm, exposure 26× 300 s, **b** expansion in He + 4 % CF_4, $\vartheta_s = 85\,°\text{C}$, $p_0 = 0.9$ bar, $p_b = 1.3\,\text{mbar}$, $d_n = 1\,\text{mm}$, exposure 12× 300 s. *Calculation* effective temperature used in single gauche to all-trans weighting = 80 K, hairpin to all-trans weighting = 0.2:1, wavenumber scaling factor = 0.99

gauche conformer contributes with a variety of bands, giving a complex pattern close to the accordion vibration band which is well reproduced by the simulation. The broad multi gauche D-LAM band spread from ≈200–250 cm^{-1} becomes more prominent. In agreement with its multi gauche origin, its intensity decreases in colder He/CF_4 expansions—an observation which is common to all low-frequency spectra. The rather broad band observed at ≈400 cm^{-1} is not covered by the simulation and assigned to multi gauche conformers accordingly.

The band located at 243 cm^{-1} is the most interesting spectral feature, because it is the first one which is finally compatible with a hairpin assignment exceeding 10 % of the all-trans abundance. A multi gauche assignment is at variance with the width of the signal, which is comparable to widths of single gauche and all-trans bands. On the other hand, the gauche 6–8 conformers contribute with weak bands in this wavenumber range, and the assignment must be reviewed carefully. The assignment actually rests upon the signals' intensity. If the hairpin conformer were present only in very low abundance, the octadecane spectrum would be expected to resemble the hexadecane and heptadecane spectra in this wavenumber range, where rather diffuse and weak bands were observed. In contrast to these bands, which could well be weak single gauche or unspecific higher gauche vibrations, the designated hairpin band stands out from the background much clearer, in line with the calculated Raman spectrum of the hairpin conformer which predicts two vibrations contributing to this band. Seemingly, this fortunate coincidence is what makes the Raman band stand out in the first place. Indeed, the single hairpin vibration predicted to show up at 312 cm^{-1} has no evident experimental counterpart. The situation can actually be reconstructed using the B3LYP simulation: setting the hairpin abundance to zero yields hexadecane- and heptadecane-like spectra (additional simulated spectrum labeled "without hairpin" in Fig. 4.15), while the match between simulation and experiment is considerably better when the hairpin abundance is increased. The single gauche abundance relative to all-trans used for simulating the octadecane spectra ranges from 5 % for gauche 8 to 16 % for gauche 5 (Table 4.4), while the hairpin abundance amounts to 20 % of the all-trans abundance ($N_{hp}/N_t = 0.2$).

As a consequence of these findings, the first positive evidence of hairpin conformers present in supersonic expansions is attributed to octadecane. However, the corresponding assignment involves just one band and is thus highly reliant on the quality of the B3LYP prediction.

Nonadecane, Eicosane, and Heneicosane

Low-frequency jet spectra change notably when the chain length is increased from octadecane to nonadecane (Fig. 4.17). A variety of new, somewhat broader and more intense bands grow from the broad D-LAM band centered at $\approx$200 cm^{-1}. Calculated single gauche abundances and scattering cross-sections demand that these conformers play a minor role in this wavenumber range and hairpin assignments are strongly supported by the B3LYP prediction. With the multiple evidence, the situation is much clearer than in the case of octadecane. The main single gauche contributions are five bands, predicted to neighbor the all-trans accordion vibration band (121 cm^{-1}) and found as very weak satellites between 125 and 175 cm^{-1} in the jet spectra. Calculated hairpin spectra involve two related hairpin conformers and are seen to agree well with the complex pattern of the jet spectra, but there are shortcomings regarding the intensities of bands above 250 cm^{-1}. The experimentally found increased width of the hairpin bands is accounted for by an increased width of Gaussian curves making up the simulation (6 cm^{-1} instead of the otherwise applied 4 cm^{-1} for all-trans and single gauche bands).

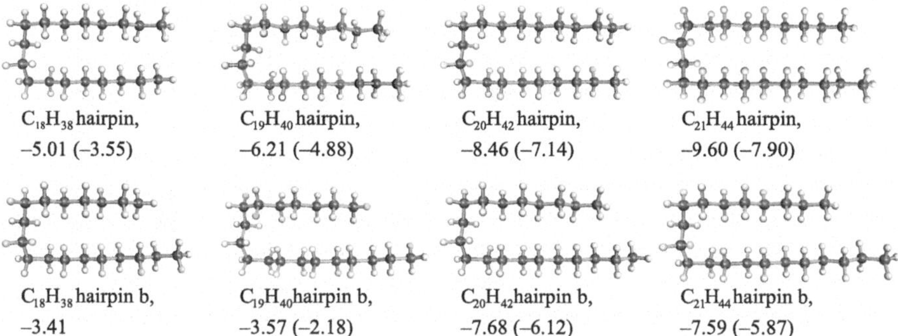

Fig. 4.16 Nomenclature of hairpin conformers. The structures were optimized on the B3LYP-D3/6-311++G** level with Turbomole v6.4 [44]. Energies relative to all-trans conformers (sum of electronic SCF energy and ZPVE) are given in kJ mol^{-1}. ZPVE corrected relative energies from B3LYP-D3/def2-TZVP optimizations/frequency calculations are given in parentheses

In the conformer denoted "hairpin b" (Fig. 4.16), the kink is one C–C segment closer to the chain end, yielding a smaller dispersion stabilization energy. In agreement with this disadvantage, the attributed abundance of the hairpin conformer b (30 %) is about half the abundance of the "optimal" hairpin conformer (50 %), which profits from a maximum of stabilizing dispersion forces. From the quantum chemical scattering strength viewpoint, both structures are more abundant than single gauche conformers (10–18 %, Table 4.4).

For eicosane, one encounters a notable deviation of predicted and observed Raman all-trans spectrum (Fig. 4.18). The eicosane accordion vibration (115 cm^{-1}) is involved in substantial mode mixing with an a_g symmetric transversal acoustic mode (110 cm^{-1}), which is exaggerated by the B3LYP calculation. This issue is of particular importance because single gauche and hairpin abundances are determined relative to all-trans, based on the accordion vibration band integral, on which the spectra are scaled. In this case the values must be thus treated with more caution. Other than that, the step from nonadecane to eicosane does not change the low-frequency jet spectra in a qualitative way, like the previous step from octadecane to nonadecane did. The bands indicative of hairpin conformers are even more distinct, but this is not reflected by the hairpin and hairpin b abundance, which amount to 30 and 50 %, respectively. However, this aspect might be due to the somewhat more uncertain all-trans integral. Also, the hairpin bands are less broad than nonadecane hairpin bands, and modeled with a FWHM of 5 cm^{-1}. The further growth of the broad and temperature sensitive D-LAM band at ≈200 cm^{-1} indicates a higher abundance of multi gauche species, demonstrating the entropic preference of these conformers with increasing chain length. Single gauche bands make up only a very minor part of the observed jet spectra, which is reflected by a seemingly even lower conformational temperature of ≈80 K, but the temperature drop could also be an artifact of the exaggerated predicted LAM-1/TAM coupling.

Heneicosane with a chain length of $n = 21$ carbon atoms was the longest alkane measured in the course of this work. Spectra in helium expansions show a massive increase of hairpin abundance compared to eicosane, but the comparability of the spectra is somewhat handicapped because of considerable air impurities in heneicosane expansions. Spectra of heneicosane expanded in He/CF_4 mixtures were not recorded. The hairpin abundance jumps to 170 % for the hairpin and 60 % for the hairpin b conformer, relative to all-trans. The match of experiment and simulation is close, but intensities from vibrations at the high-frequency flank of the D-LAM band deviate notably once again. Single gauche bands almost completely vanish into the background noise so that the conformational temperature of 90 K is rather an upper bound.

Comment: Hairpin skeletal vibrations

So far, low-frequency vibrational bands were assigned to specific conformers, but the vibrational character was hardly discussed beyond the statement that these vibrations are similar to the accordion vibration. A detailed discussion of single gauche vibrations would not add to the discussion of the hairpin/all-trans competition, but hairpin vibrations are interesting because there is an underlying classification scheme which reinforces the assignment and even leads to the assignment of another heneicosane hairpin conformer. When examining assigned hairpin vibrations more closely, one finds groups of vibrations with similar wavenumbers and intensities and even vibrations which bear virtually the same wavenumber in the present low-resolution spectra. By comparison to calculated normal modes from the simulated spectra, this is seen to derive from common vibrational displacements, all of which involve a longitudinal accordion-like stretching of one or both hydrocarbon hairpin arms, but different transversal movement and twisting. Among the assigned hairpin vibrations, four groups can be distinguished, which are termed type I–IV in the following and depicted in Fig. 4.20, using the example of nonadecane. The classification is listed in Table 4.8 and further normal coordinates are visually compared in Appendix A.5.

Type I and II vibrations are quite similar. The longitudinal arm stretching goes together with transversal methylene displacement and twisting. The lower-wavenumber type I vibrations (170–210 cm^{-1}) contrast with type II vibrations (210–240 cm^{-1}) by the higher amplitude in transversal displacement and twisting, which takes place in several methylene units close to the kink, while this movement is limited to methylene units directly adjacent to the kink in type II vibrations. Type III vibrations are longitudinal stretches of short arms, in agreement with a higher wavenumber between 250 and 270 cm^{-1}, where methylene twisting is restricted to the central –CH_2– units of the kink. Type IV vibrations are at higher wavenumbers near 290–310 cm^{-1}. They involve longitudinal stretching of both arms with mid-arm nodes, in phase with a compression of the kink by transversal displacement and twisting. This sets type IV vibrations apart from type I and II vibrations, where the phase relation of longitudinal and transversal displacement leads to an expansion of the kink when the arms are stretched. Type II vibrations bear the highest Raman scattering intensity, followed by type I, and type III and IV which are similar in this aspect.

Table 4.7 Assignment of experimental low-frequency Raman bands: nonadecane, eicosane, and heneicosane

Nonadecane		Eicosane		Heneicosane	
Wavenumber	Assignment	Wavenumber	Assignment	Wavenumber	Assignment
121 (s)	All-trans[a]	110 (w,sh)	All-trans[b]	109 (m)	All-trans[a]
129 (vw)	Gauche 4, *1, 2*	115 (s)	All-trans[a]	169 (w)	Hairpin b
136 (vw)	*Gauche 2, 5*	126 (vw)	Gauche 2, 4 ,*5*	192 (m)	Hairpin
143 (vw)	Gauche 3, *8*	150 (vw)	Gauche 7	204 (m)	Hairpin
153 (vw)	Gauche 7	170 (vw)	Gauche 5, 6, *4, 8*	209 (w,sh)	Hairpin b
161 (vw)	*Gauche 6*	193 (vw)	*Hairpin b*	221 (m)	Hairpin
175 (vw,b)	*Gauche 5, 6*	203 (w)	Hairpin	250 (vw)	Hairpin b,
193 (vw)	*Hairpin b,*	222 (w)	Hairpin b	268 (vw)	*Hairpin c*[d]
	Gauche 8	229 (vw)	Hairpin	292 (w)	Hairpin,
203 (w)	Hairpin,	252 (w)	*Hairpin b*		Hairpin b
	Hairpin b	294 (vw)	Hairpin	310 (w,sh)	All-trans[c]
220 (vw)	Hairpin b	306 (vw)	Hairpin b		
233 (w)	Hairpin	320 (w)	All-trans[c]		
249 (vw)	Hairpin				
267 (vw)	Hairpin b				
293 (w)	Hairpin				
311 (w,b)	Hairpin,				
	Hairpin b				
340 (m)	All-trans[c]				

Tentative assignments are printed in italics. Wavenumbers are given in cm^{-1}, relative intensities in parentheses (b = broad, v = very, w = weak, m = medium, s = strong). The number characterizing gauche conformers indicates the position of the gauche bond (see Fig. 4.9)

[a]LAM-1; [b]TAM with LAM-1 character; [c]LAM-3; [d] see discussion on p. 84

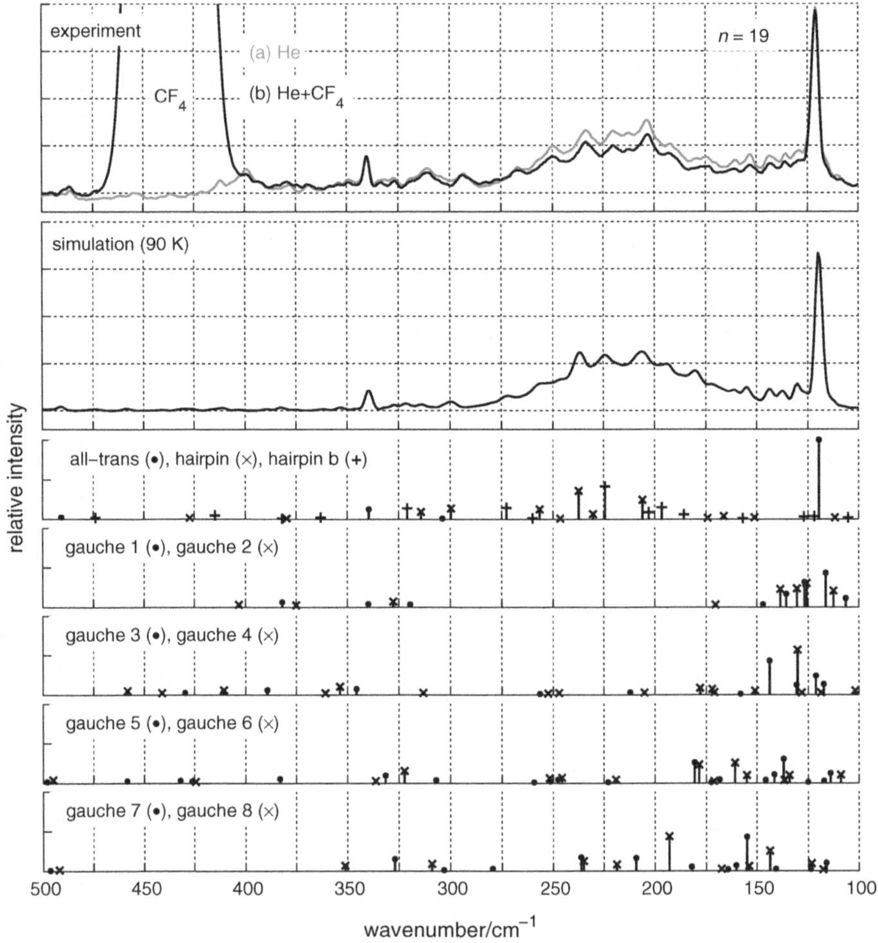

Fig. 4.17 Comparison of averaged jet-cooled Raman spectra of nonadecane in the low-frequency region to a B3LYP-D3/6-311++G** simulation. Experimental spectra are Savitzky-Golay filtered (9 pt.). *Measurement conditions* **a** expansion in He, $\vartheta_s = 95$ °C, $p_0 = 0.5$ bar, $p_b = 0.9$ mbar, $d_n = 1$ mm, exposure 16× 300 s, **b** expansion in He + 4% CF_4, $\vartheta_s = 95$ °C, $p_0 = 0.9$ bar, $p_b = 1.3$ mbar, $d_n = 1$ mm, exposure 8× 300 s. *Calculation* effective temperature used in single gauche to all-trans weighting = 90 K, hairpin (hairpin b) to all-trans weighting = 0.5 (0.3):1, wavenumber scaling factor = 0.99, FWHM hairpin = 6 cm^{-1}

The wavenumbers of these accordion-like hairpin vibrations are connected to the segment length of longitudinally moving arms much like accordion vibrations of all-trans conformers are connected to the overall chain length [42], but it is somewhat unclear which methylene segments must be apportioned to the segment length, or "effective chain length" (n'), of the longitudinal modes. However, there are two cases where arms of the same length vibrate and counting is of no further importance.

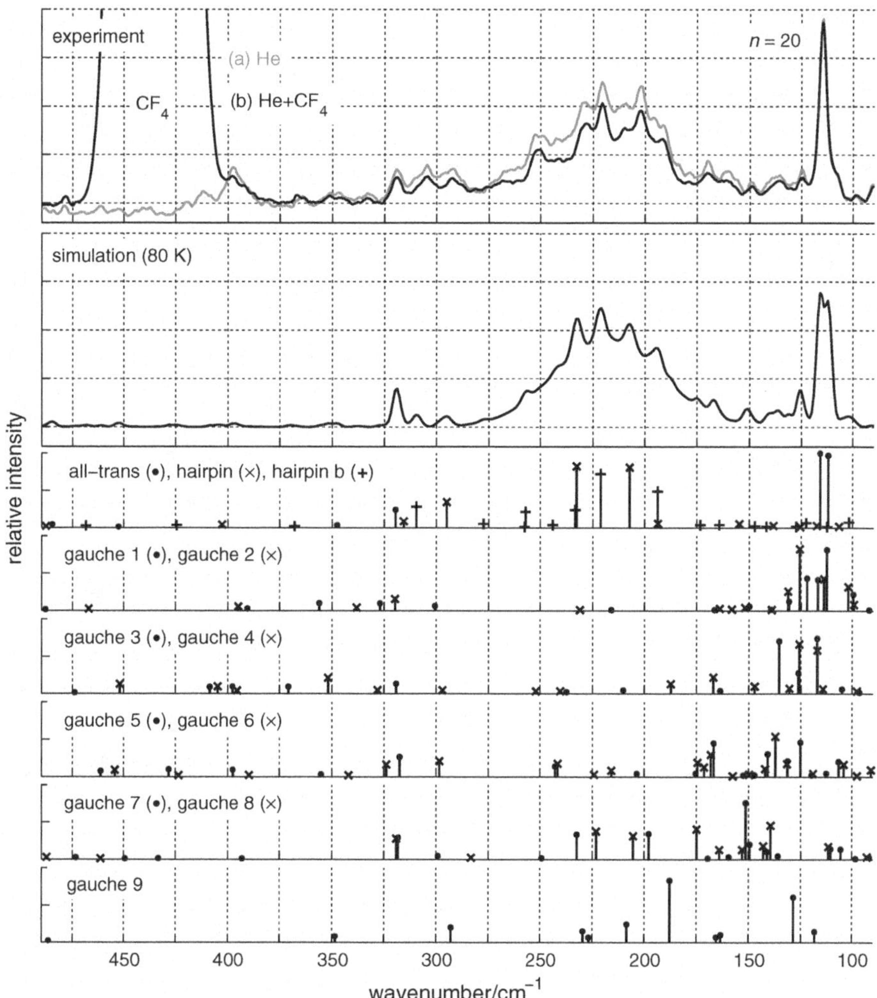

Fig. 4.18 Comparison of averaged jet-cooled Raman spectra of eicosane in the low-frequency region to a B3LYP-D3/6-311++G** simulation. Experimental spectra are Savitzky-Golay filtered (9 pt.). *Measurement conditions* **a** expansion in He, $\vartheta_s = 105°C$, $p_0 = 0.5$ bar, $p_b = 0.9$ mbar, $d_n = 1$ mm, exposure 8× 300 s, **b** expansion in He + 4% CF_4, $\vartheta_s = 105°C$, $p_0 = 0.9$ bar, $d_n = 1$ mm, exposure 8× 300 s. *Calculation* effective temperature used in single gauche to all-trans weighting = 80 K, hairpin (hairpin b) to all-trans weighting = 0.3 (0.5):1, wavenumber scaling factor = 0.99, FWHM hairpin = 5 cm^{-1}

The first case are type I vibrations for $n = 19$–21 which share virtually the same wavenumber (203, 203, and 204 cm^{-1}, respectively). The corresponding predicted normal modes show that the heneicosane vibration involves only the short arm in the longitudinal displacement, the eicosane vibration both arms, and the nonadecane vibration only the long arm. In each case, the longitudinally moving segments are of

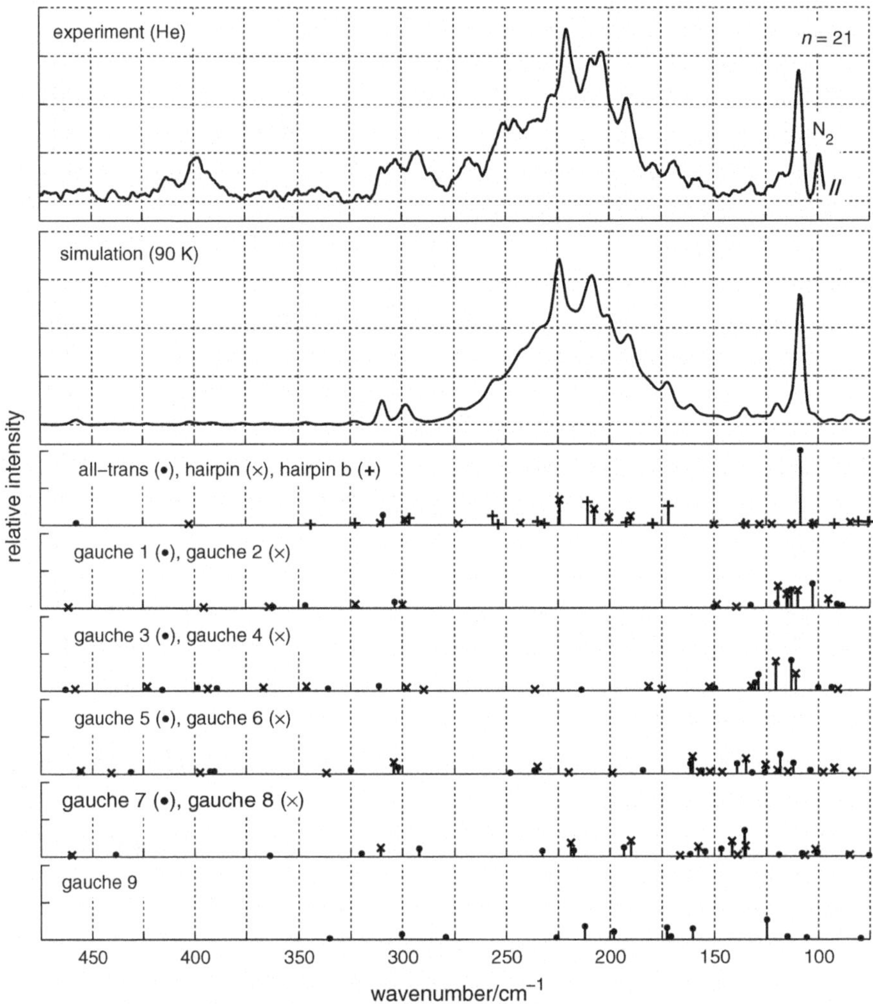

Fig. 4.19 Comparison of an averaged jet-cooled Raman spectrum of heneicosane to a B3LYP-D3/6-311++G** simulation in the low-frequency region. The experimental spectrum is Savitzky-Golay filtered (9 pt.). *Measurement conditions* expansion in He, ϑ_s = 125 °C, p_0 = 0.55 bar, p_b = 0.8 mbar, d_n = 1 mm, exposure 8× 600 s *Calculation* effective temperature used in single gauche to all-trans weighting = 90 K, hairpin (hairpin b) to all-trans weighting = 1.7 (0.6):1, wavenumber scaling factor = 0.99, FWHM hairpin = 5 cm^{-1}

the same length, yielding a nearly perfect wavenumber match. The same is true for type II vibrations of heneicosane (221 cm^{-1}), eicosane (hairpin b, 222 cm^{-1}), and nonadecane (hairpin b, 220 cm^{-1}), where the long arms, all of which of the same length, are longitudinally stretched and clinched along the normal coordinate. A way to check whether there is a systematic dependency for the remaining vibrations is to

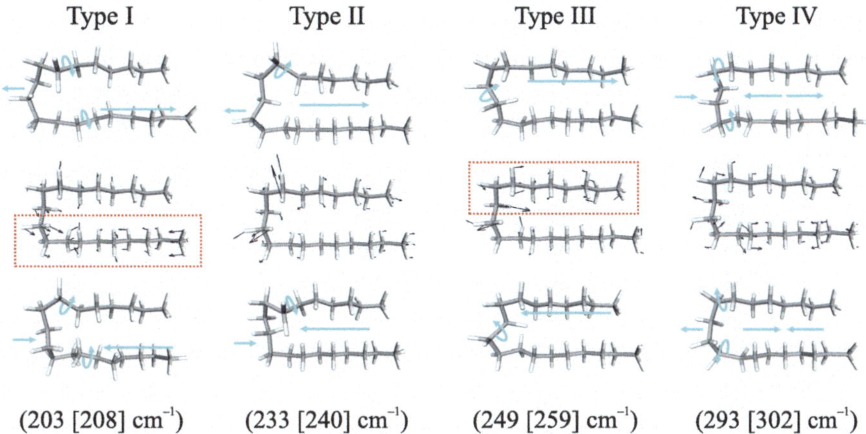

(203 [208] cm^{-1}) (233 [240] cm^{-1}) (249 [259] cm^{-1}) (293 [302] cm^{-1})

Fig. 4.20 Exemplary skeletal hairpin vibrations of nonadecane. Normal coordinates are shown in the middle, framed by exaggerated turning points below and above. *Blue arrows* illustrate the movements along the normal coordinates qualitatively, *red boxes* indicate segments counted as the effective chain length (both arms participate in the vibration at 233 and 293 cm^{-1}, and the effective chain length is counted as half the overall chain length). Vibrational wavenumbers are given at the *bottom* (unscaled B3LYP-D3/6-311++G** calculation in square brackets)

count the longitudinally moving segments or effective chain length and plot it against the wavenumber of the corresponding vibration. This should lead to a hyperbolic dependency like reported by Shimanouchi for LAMs of normal alkanes [34], which will be discussed in Sect. 6. The question what segments should be counted can be approached by comparing to work from Grossmann and Bölstler who published spectra of cycloalkanes [74, 75]. Cycloalkanes are known to assume the structure of a "collapsed ring" in the solid state [76], close to the structure of hairpin conformers. Grossmann and Bölstler found that LAMs of cycloalkanes $(CH_2)_n$ oscillate with wavenumbers of LAMs from normal alkane with half the chain length $C_{n/2}H_{n+2}$, meaning that the kink does not affect the position of the LAM and moves as a rigid mass. From the calculated normal coordinates one can see that this is not the case for hairpin conformers, but it is a useful first approximation. Accordingly, if both arms are displaced in a longitudinal fashion, half the chain length is taken as the effective chain length, otherwise the moving arm is counted together with half the kink (indicated by red boxes in Fig. 4.20).

The resulting plot is depicted in Fig. 4.21. Indeed, the Figure shows a systematic development for type II and III vibrations, reasonably close to Shimanouchi's approximation $\tilde{\nu}/\text{cm}^{-1} = 2400/n$, but no such trend for type I and IV vibrations. These findings actually invite for a more sophisticated mechanical analysis, but this is beyond the scope of this work. Instead, the findings should underline the overall consistent picture which results from the hairpin assignments based on the B3LYP prediction. Not only is the match of simulation and experiment very close, but the assignment makes also sense from a mechanical point of view.

Table 4.8 Characterization of skeletal hairpin vibrations

Chain length	Stretching arm(s)	Exp. wavenumber	Calc. wavenumber[a]	Scattering cross section
Type I: twist of several CH_2 close to kink (away from molecule when arms are stretched, kink expanded)				
21 b	both	172	174	134
20 b	long	193	196	109
19	long	203	208	102
20	both	203	209	179
21	short	204	209	111
Type II: twist of CH_2 adjacent to kink (away from molecule when arms are stretched, kink expanded)				
21 b	long	209	213	162
19 b	long	220	227	171
21	long	221	227	177
20 b	long	222	223	161
20	both	229	235	183
19	both	233	240	147
Type III: twist of central kink CH_2 against stretch of short arm				
19	short	249	259	51
21 b	short	250	259	64
19 b	short	267	276	58
21 c	short	268	N/A	N/A
Type IV: twist of kink CH_2 (towards molecule when arms are stretched, kink compressed)				
21 b	both	292	299	55
21	both	292	302	41
19	both	293	302	58
20	both	294	298	79
20 b	both	306	313	68
19	both	311	317	52
19 b	both	311	324	60

The addition b (c) on the chain length indicates less stabilized hairpin conformers (see Fig. 4.16). Wavenumbers are given in cm^{-1}, scattering cross-sections in $10^{-36} m^2 sr^{-1}$ (calculated for $T = 100$ K). The calculation level is B3LYP-D3/6-311++G** (Turbomole v6.4 [44])
[a] Unscaled

The above discussion invites for an additional hairpin assignment in case of heneicosane: the 268 cm^{-1} band could stem from a weak vibration belonging to the common hairpin conformer. However, according to the calculated intensity it should be less prominent in this case. An alternative is to assign it to a type III longitudinal short arm stretch vibration of another hairpin-type conformer, judging from the nonadecane hairpin b band at 267 cm^{-1}. The corresponding heneicosane conformer with the same short arm chain length is a hairpin with the kink located even one

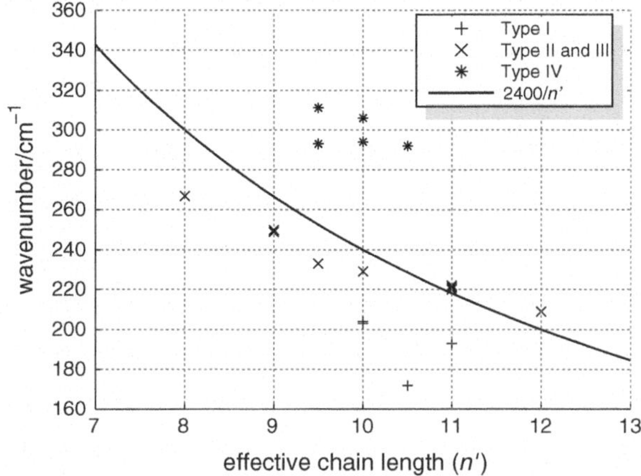

Fig. 4.21 Skeletal hairpin vibration wavenumbers plotted against the effective chain length (n'). The *solid line* indicates Shimanouchi's approximation for normal alkane LAM-1 vibrations ($\tilde{\nu}/\text{cm}^{-1} = 2400/n'$) [34]

bond closer to the chain end compared to the heneicosane hairpin b conformer—hence termed hairpin c (Tables 4.7 and 4.8). The hairpin c type II long arm vibration (effective chain length $n' = 13$) should be located at $\approx$200 cm^{-1}, extrapolating from the plot in Fig. 4.21, which cannot be verified due to hairpin and hairpin b overlap. An additional indication that there are more hairpin-type conformers contributing to the spectrum is the band at 303 cm^{-1}, which is left unassigned. The hairpin c assignment could be validated by further quantum chemical calculations.

Spectral development with chain length

An overview of the Raman jet spectra recorded in the low-frequency region from helium expansions is arranged in Figs. 4.22 and 4.23. The variation of the signal-to-noise ratio stems partly from the number of spectra available for each alkane and partly from fluctuations in concentration and sensitivity of the Raman spectrometer. The decane, undecane and dodecane spectra were not discussed before and are shown merely for a more complete picture. They were measured as test cases to compare the conventional glass saturator, used to prepare gas mixtures at temperatures below room temperature, and the heatable brass saturator (Sect. 3.1).

The spectra show some apparent trends correlated with the chain length. Focusing on the accordion vibration, which is usually the most intense sharp band, one finds the known proportionality of the accordion vibration wavenumber to the inverse of the chain length [34] which results from the steep and essentially linear increase of the C–C–C bending frequency branch at low phase values and the decreasing phase value of the $m = 1$ LAM with increasing chain length (see Chap. 6). Another striking development is the relative intensity of Raman bands. From decane to pentadecane the accordion vibration band is accompanied by a varying number of bands which make up a less but still important contribution to the overall spectrum, whereas the intensity

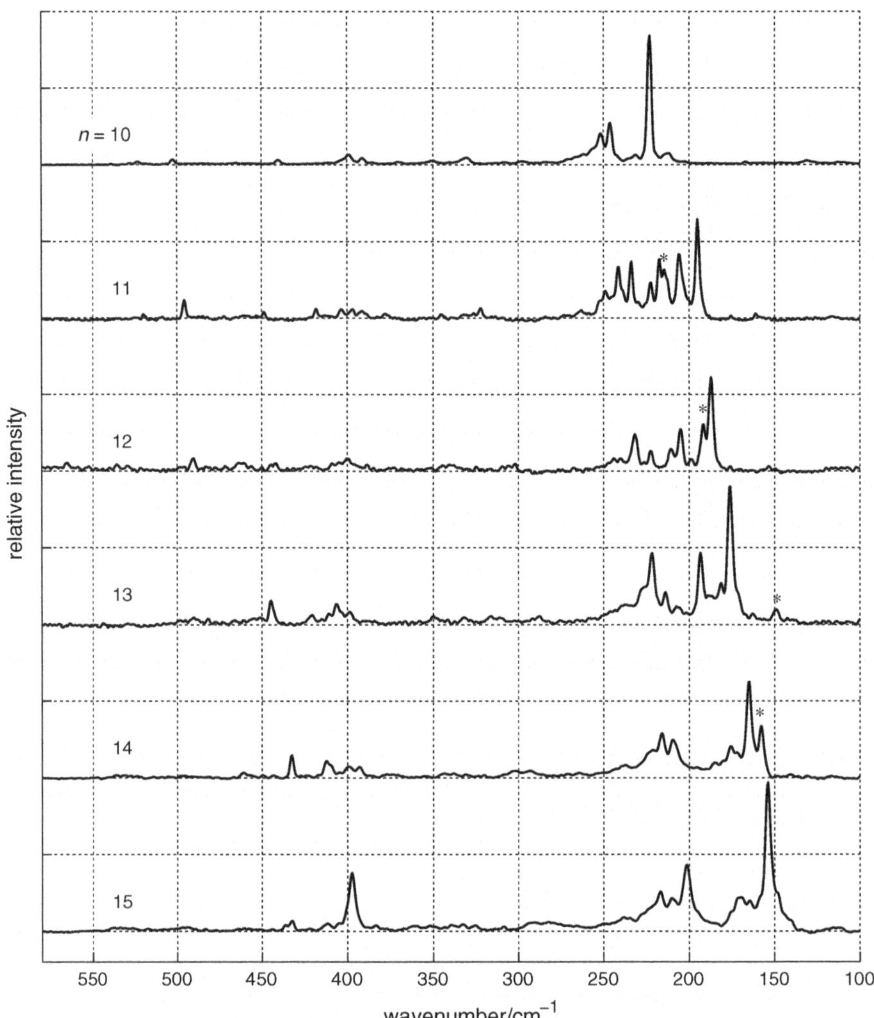

Fig. 4.22 Averaged jet-cooled Raman spectra of the accordion vibration region of alkanes with chain length n in He expansions. Spectra are scaled to the overall accordion peak intensity including signals of coupling vibrations ($n = 11, 13$: harmonic mode mixing, $n = 12, 14$: anharmonic Fermi resonance, marked with $*$). Reference II—Reproduced by permission of The Royal Society of Chemistry

of such satellite bands drops notably at $n = 16$. Close to the accordion vibration, these signals are often identified as accordion-like vibrations from single gauche conformers, but also occasional harmonic mode mixing ($n = 11, 13, 18$ and 20) and Fermi resonances ($n = 12$ and 14) play an important role by redistributing intensity of the all-trans accordion vibration (marked with $*$ in the spectral overview). Undecane, $n = 11$, is an outstanding example for strong harmonic mode mixing, and tetradecane

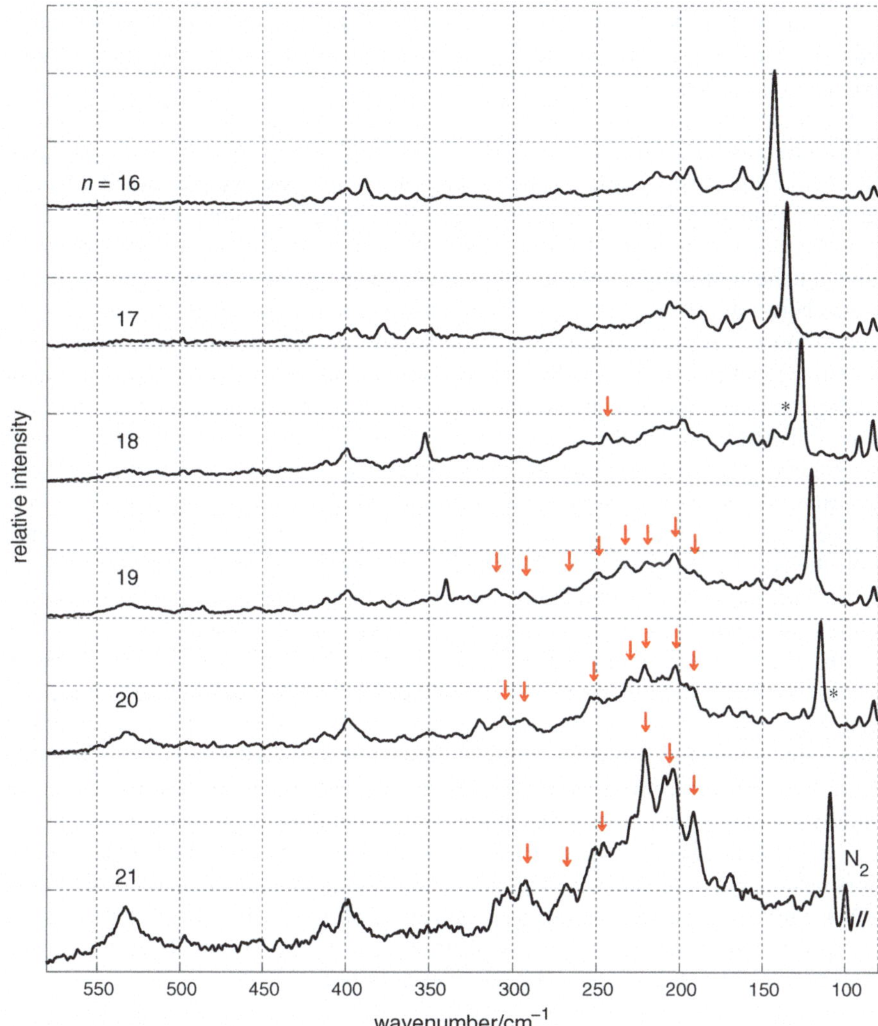

Fig. 4.23 Averaged jet-cooled Raman spectra of the accordion vibration region of alkanes with chain length n in He expansions (see Fig. 4.22). Mode mixing is considered in case of $n = 18, 20$ (marked with $*$). Bands assigned to hairpin conformers are marked with *arrows*. Air impurities ($\tilde{\nu} \leq 100\,\mathrm{cm}^{-1}$) are prominent especially in the case of $n = 21$. Reference II—Reproduced by permission of The Royal Society of Chemistry

for a strong Fermi resonance [6]. The relative intensity drop of single gauche bands for $n \geq 16$ is seemingly due to a lower effective conformational temperature (in the range 100–110 K for $n = 13$–15 and 80–90 K for $n = 16$–21), but also because single gauche bands are scattered more strongly at higher chain length and accidental band overlap happens less often. A prominent single gauche band overlap was seen to explain the high Raman scattering intensity of the pentadecane LAM-3 band at 398 cm^{-1}, which is at variance with the LAM-3 band intensity of the other alkanes (discussed on p. 70).

The increasing intensity of broad bands indicates the growing number of accessible multi gauche conformers at higher chain length. For tridecane to pentadecane the LAM-1 and close by single gauche bands are seen to sit on top of broader bands, but their change in Raman intensity with chain length is rather small. At higher chain length, the broader D-LAM band [38, 39], a superposition of bands from multi gauche conformers, first slowly and then rapidly grows from the spectral background between 150 and 250 cm^{-1}. It demonstrates how the conformational diversity becomes more pronounced at higher chain length and that the supersonic expansion technique reaches a limit regarding its conformational cooling efficiency. Especially the step from eicosane to heneicosane comes along with a distinct intensity increase of the D-LAM band, but the effective conformational temperature might be affected by a higher concentration of air impurities in case of heneicosane. The heneicosane spectrum also helps to clearly identify broad and chain length independent bands at $\approx$400 and $\approx$540 cm^{-1}, which show the same behavior as the D-LAM band and may likewise be assigned to multi gauche conformers, adding to an overall more liquid character of the spectra at high chain length. However, only the former signal is found in liquid phase *n*-alkane spectra [5], while the latter band at $\approx$540 cm^{-1} is not observed. On the other hand, Raman spectra of cycloalkanes [37, 74, 75] with structures related to the hairpin conformer (see above) show bands close to this position, suggesting a connection to hairpin conformers. The lack of a corresponding band in the simulated spectrum could be explained by an anharmonic origin, or an origin from hairpin conformers with gauche defects in the aligned arms. The most important evolution of the hairpin contribution to the Raman spectra can be summarized as being negligible for alkanes shorter than octadecane and small but detectable for octadecane. For nonadecane and eicosane the hairpin abundance is considerably higher with evidence for two different hairpin conformers. In case of heneicosane, the hairpin conformer outweighs the all-trans conformer. The implications for the all-trans/hairpin concurrence will be discussed in Sect. 4.5.

4.4.2 C–C Stretching Region

In the spectral region from 800 to 1400 cm^{-1}, *n*-alkanes possess strongly Raman active C–C stretching and C–H deformation vibrations. For the characterization of these vibrations, reference to polyethylene dispersion curves becomes helpful. In contrast to the low-frequency skeletal vibrations, the vibrations in this region are much less sensitive to the chain length and conformation and the evolution with increasing chain length involves more subtle changes. For this reason, the assignment does not address each single chain length separately, but starts with an overview including all alkanes measured in this wavenumber range (Fig. 4.24) and corresponding simulations (Fig. 4.25). To simulate the spectra, a uniform conformational temperature (effective single gauche temperature) of 100 K was chosen for better comparability. Matching experimental intensities to deduce the conformational temperature is not possible because of significant multi gauche contributions. As examples for short, medium sized, and long alkanes, more detailed presentations of spectra and simulations for tridecane, heptadecane, and heneicosane are included in the Appendix A.9 (Figs. A.17–A.19).

The findings from the low-frequency region, namely that the hairpin conformer contributes significantly to the Raman spectra for $n > 18$, enables one to identify possible hairpin bands by their behavior, without a detailed assignment. For quick reference, the spectra are divided in three regions, centered at ≈900, 1100, and 1300 cm^{-1}. Major changes with chain length are seen to occur in the 1100 cm^{-1} region, followed by 900 cm^{-1}, while the band at ≈1300 cm^{-1} remains rather unchanged. In contrast to the many related bands observed for all chain lengths and with wavenumbers largely independent from the chain length, several bands stand out because they are either observable only at higher chain length or show a conspicuous wavenumber/chain length dependence. One band which moves in a small wavenumber interval starts at 1109 cm^{-1} for $n = 13$ and shifts to 1124 cm^{-1} for $n = 21$. The same is true for a band first clearly observed for $n = 18$ at 1087 cm^{-1} which shifts to 1094 cm^{-1} for $n = 21$. Signals related in such a way that they first appear at higher chain length are observed at 887 cm^{-1} ($n = 21$), 1111–1115 cm^{-1} ($n = 19$–21), and 1144 cm^{-1} ($n = 19$–21). Because the hairpin conformer contributes substantially to the Raman spectrum for $n > 18$, these "new" signals are hot candidates for hairpin-specific bands. The shifting band however, observed between 1087–1094 cm^{-1}, can be removed from the list of designated hairpin bands, because there is evidence that it is already present (but largely covered) in spectra for $n = 13$–17: the tridecane spectrum shows that the band at 1066 cm^{-1} has a high-frequency shoulder, which shifts towards higher wavenumber when going to the next longer alkane. In the hexadecane spectrum (the pentadecane spectrum was not recorded), the high-frequency shoulder of the 1066 cm^{-1} band has vanished, but it appears at the low-frequency slope of the next higher-wavenumber band at 1082 cm^{-1}. The 1082 cm^{-1} band of heptadecane is noticeably broader than its counterparts from other chain lengths and at octadecane, the signal finally clearly sticks out. The pentadecane spectrum would be needed to resolve this issue, but likely, it shows a band in between the 1066 and 1082 cm^{-1} bands, underlining that the moving signal is not a feature of the hairpin conformer. Thus, this first discussion leaves the signals at 887, 1111–1115, and 1144 cm^{-1} as possible hairpin features.

To clarify the origin of the signals, a detailed assignment is necessary. Fortunately, the spectral assignment in this wavenumber range is well established in the literature, down to bands specific for gauche/trans sequences [24, 77–79], including the tight fold ggtgg sequences sought in this work [37]. Assignments were collected and summarized recently by Brambilla and Zerbi (Ref. [25] and references therein) and Orendorff et al. [80]. At the high-frequency end of the spectra (≈1300 cm^{-1}), the complete in-phase CH_2 twisting vibration (τCH_2) is found as a relatively strong and broad band. Its shape becomes more asymmetric from low to high chain length while its peak wavenumber remains quite constant [29]. In agreement with the broad

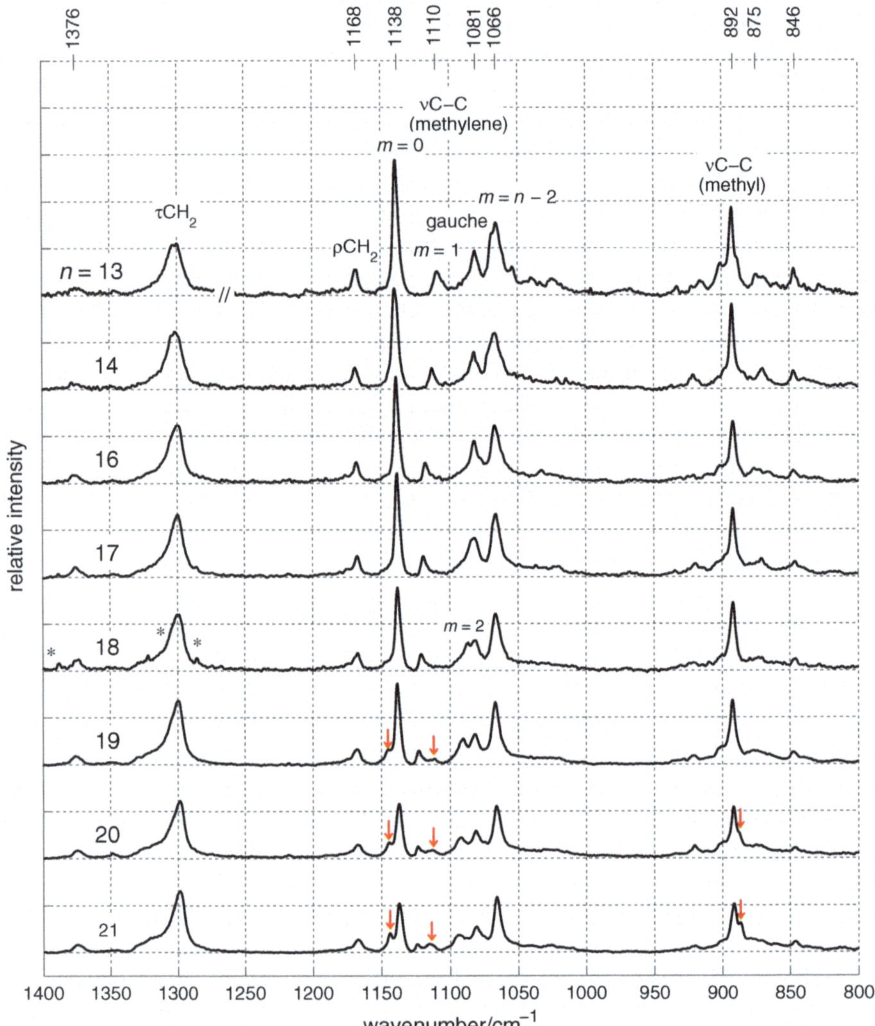

Fig. 4.24 Jet-cooled Raman spectra of the C–C stretching region of alkanes with chain length n in He expansions. Bands assigned to hairpin conformers are marked with *arrows*. The spectra are scaled to the integral of the CH_2 twisting vibration (1240–1340 cm^{-1}) and Savitzky-Golay filtered (7 pt.). In the octadecane spectrum, impurities are marked with $*$ (removed from the CH_2 twisting vibration integral *Measurement conditions* exposure 6× 600 s, otherwise analogous to low-frequency He expansion measurements. Reference II—Reproduced by permission of The Royal Society of Chemistry

and growing D-LAM band in the low-frequency region, the increasing asymmetry reflects increasing conformational disorder. This behavior is known from a Raman spectroscopic study of polyethylene which found the rather sharp part of the band

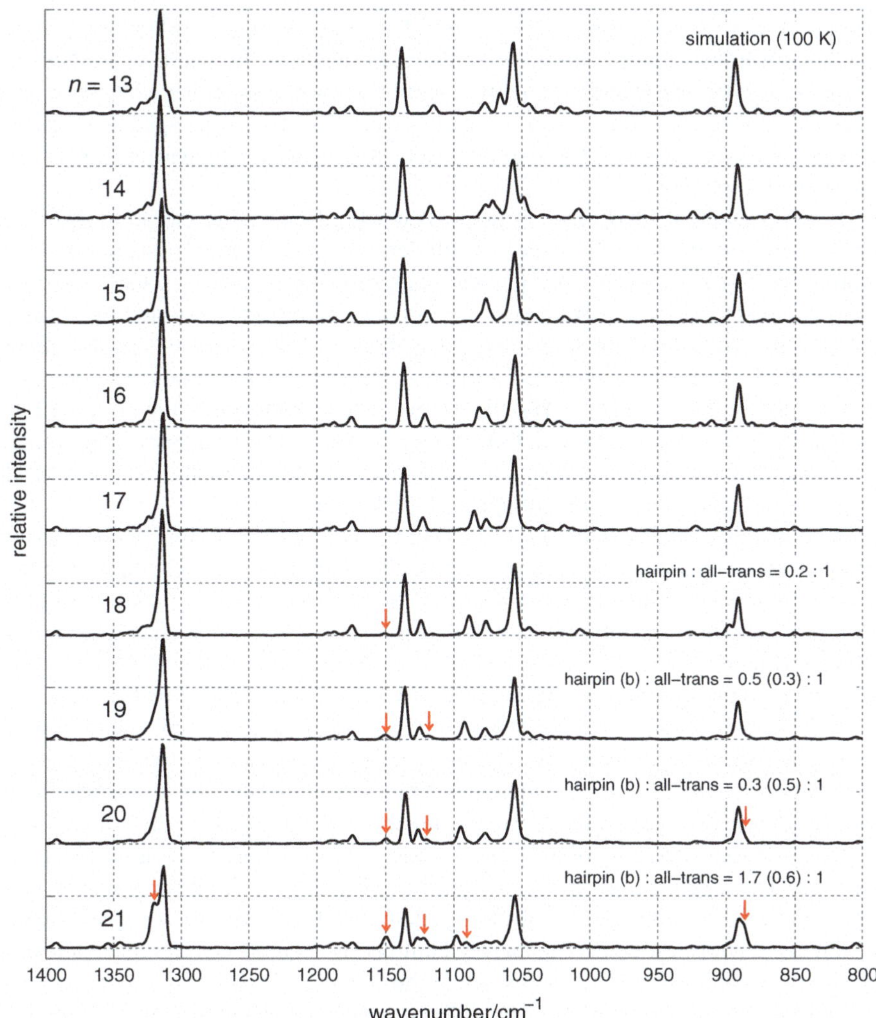

Fig. 4.25 Simulated Raman spectra from harmonic B3LYP-D3/6-311++G** calculations, see low-frequency spectra and simulations in Sect. 4.4.1. Bands which stem from hairpin conformers are marked with *arrows*. The spectra are scaled on the integral of the CH_2 twisting vibration near $1300\,cm^{-1}$, analogous to the experimental spectra. Effective temperature used in single gauche to all-trans weighting =100 K, hairpin weighting as deduced from low frequency spectra (given on the *right side*), wavenumber scaling factor = 0.99. Reference II—Reproduced by permission of The Royal Society of Chemistry

to be associated with crystalline polyethylene, mainly in the all-trans conformation, while the broad, high-wavenumber foot of the band is observed solely for conformationally disordered molten polyethylene [81]. The same study found that the integrated intensity of this band is not affected by conformational disorder. Because of this property, the τCH_2 integral was used to scale the spectra. The single conformer stem spectra presented for $n = 13$, 17 and 21 in Figs. A.17–A.19 show that single gauche conformers contribute rather to the main peak of the band, while hairpin conformers give rise to bands at slightly higher wavenumbers, either contributing to the broad foot or to a shoulder of the main peak. The Raman jet spectra show no outstanding shape for $n > 18$ so that no conclusion regarding the hairpin conformer is possible for τCH_2.

The weak band at 1167 cm^{-1} is associated with the twisting vibration. It is the opposite limiting mode of the same polyethylene frequency branch [34] (methylene twisting-rocking modes [29]). A complete change in vibrational character from rocking to twisting when going from $\varphi = 0$ to $\varphi = \pi$[12] was demonstrated by Tasumi et al. [31] in terms of the potential energy distribution [82]. Indeed, the displacement along B3LYP normal coordinates of all-trans conformers corresponds to a complete in-phase methylene rocking (ρCH_2). For single gauche conformers however, the normal coordinate involves substantial methylene twisting. The frequency is not much affected among different single gauche conformers, reflected by the observed band which is sharp and flattens only at higher chain length.

The cluster of bands centered at $\approx$1100 cm^{-1} is of particular interest because it involves two of the designated hairpin bands. The dominating signals stem from the most in-phase C–C stretching vibration at 1138 cm^{-1} (νC–C, $m = 0$) and most out-of-phase C–C stretching vibration at $\approx$1066 cm^{-1} (νC–C, $m = n - 2$). These modes differ considerably in their coupling to internal coordinates other than C–C stretching: the in-phase vibration is strongly coupled to C–C–C bending and methyl bending while the out-of-phase mode couples strongly to methylene wagging with almost no coupling to methyl vibrations. The intensity of these bands is found to behave differently with increasing chain length. While the in-phase C–C stretching band intensity clearly decreases relatively to the integral of the in-phase methylene twisting band, a systematic decrease of the out-of-phase C–C stretching band intensity is not observed. The single conformer stem spectra in Figs. A.17–A.19 show that in fact bands scattered around the out-of-phase stretching band are closely spaced, and the Raman scattering intensity is largely confined to a small wavenumber interval, especially in the limit of long chains, while the spacing of bands close to the $m = 0$ in-phase stretching band is wider. The different spacings related to in-phase and out-of-phase C–C stretching are in agreement with the polyethylene dispersion curve for C–C stretching, which is rather flat at high phase values (out-of-phase vibrations) and more steep close to low phase values (in-phase vibrations) [34].

The Raman band between 1109 and 1124 cm^{-1} was discussed above as one of the signals shifting with increasing chain length. Referring to the B3LYP simulation and the literature [35], it is assigned to the C–C stretching vibration with one node ($m = 1$). This assignment explains the signal's chain length dependence, because one node translates to a different phase value for each chain length. Since the polyethylene C–C stretching frequency branch drops when going from $\varphi = 0$ to small $\varphi > 0$ values [29], and the phase associated with $m = 1$ decreases with increasing chain

[12] Snyder uses the opposite phase description for the methylene twisting-rocking modes, because of a different coordinate definition [29].

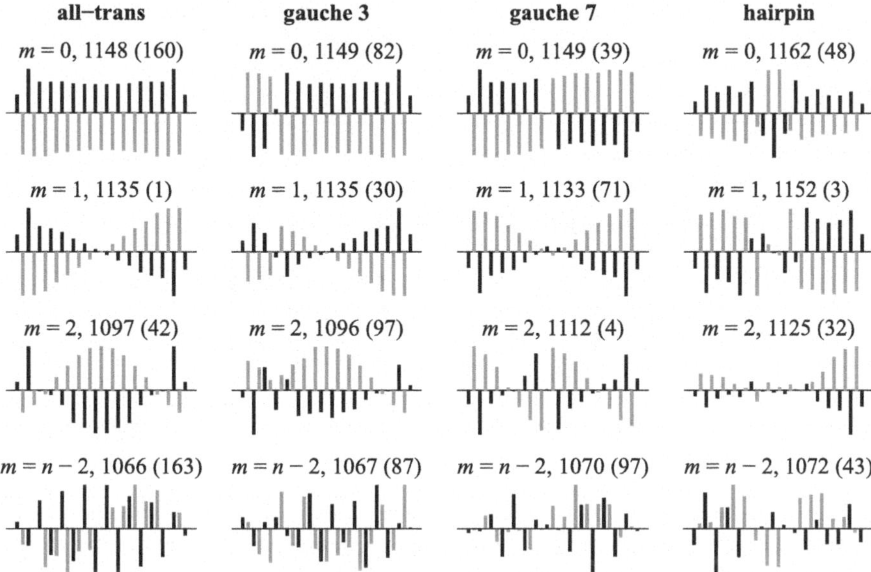

Fig. 4.26 Internal coordinate displacements of C–C stretching normal vibrations using the examples of representative heptadecane conformers, calculated on the B3LYP-D3/6-311++G** level. *Black bars* indicate relative C–C stretching amplitudes, *gray bars* relative C–C–C bending amplitudes (see Sect. 4.2, Fig. 4.6). Numbers include unscaled harmonic wavenumbers in cm^{-1} and differential scattering cross sections (in parentheses, $T_{vib} = 100\,K$) in $10^{-36}\ m^2\,sr^{-1}$

length, the corresponding vibration must shift from lower frequencies towards the higher-frequency $m = 0$ vibration, as found in the experiment.

The $m = 1$ mode is fairly Raman active in case of single gauche conformers but weak or symmetry-forbidden in case of all-trans conformers. The redistribution of intensity from the $m = 0$ to the $m = 1$ mode for gauche conformers can be understood by examining the corresponding normal coordinates, illustrated in Fig. 4.26 for gauche 3 and 7 heptadecane. The phase of C–C stretching oscillators switches sign at the gauche bond [35]. Therefore, for $m = 0$ vibrations, local polarizability derivatives do not add up as efficiently as it is the case for the all-trans conformer and the Raman scattering intensity decreases. For gauche $m = 1$ vibrations, local polarizability derivatives add up, and they do so most efficiently when the node of the vibration and the gauche bond coincide, that is, when the gauche bond is as close as possible to the chain center (gauche 7 in case of the heptadecane example). While the intensity is quite sensitive to the presence of a gauche bond, the frequency is seen to be less sensitive. The $m = 1$ C–C stretching band decreases in intensity quite uniformly along with the $m = 0$ band, with peak intensity ratios scattering around $\approx$5–7. Therefore, double and higher gauche conformers, which are more abundant in case of longer alkanes do not seem to contribute to this band as much as single gauche does, and the main intensity contribution should stem from single gauche conformers with the gauche conformation close to the chain center.

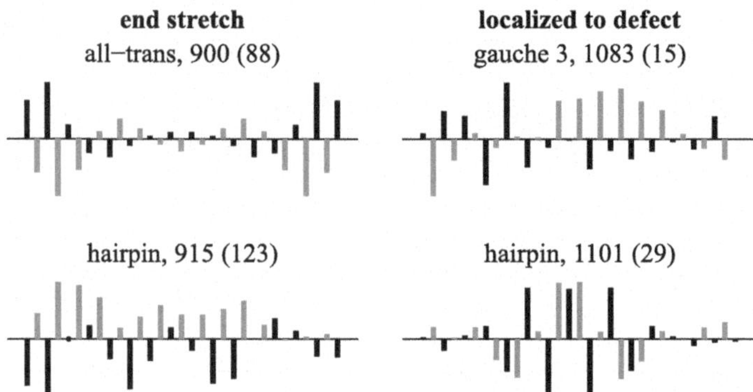

Fig. 4.27 Internal coordinate displacements of heptadecane C–C stretching normal vibrations, see Fig. 4.26. The depicted example vibrations are localized to the chain ends, gauche bonds, or the hairpin ggtgg sequence

The other band which shifts with increasing chain length was also mentioned above and is clearly observable only in case of $n \geq 18$. It is assigned to the $m = 2$ C–C stretching vibration [29], also illustrated in Fig. 4.26. The wavenumber/chain length dependence is thus explained in analogy to the $m = 1$ vibration by the dispersion of the C–C stretching frequency branch. In the simulated series of spectra shown in Fig. 4.25 the $m = 2$ vibration is found first at the high-frequency slope of the $n = 13$ out-of-phase C–C stretching vibration and moves stepwise up to $\approx$1100 cm^{-1} for $n = 21$, which is the behavior indicated by the jet-spectra. The assignment to a hairpin-specific feature can be firmly rejected.

At $\approx$1080 cm^{-1} one finds C–C stretching vibrations localized in gauche bonds [79]. The localization is apparent when examining internal coordinate displacements from the B3LYP calculations, shown in Fig. 4.27 for a gauche 3 heptadecane vibration. The band is thus missing in Raman spectra of crystalline all-trans alkanes, but of high intensity in Raman spectra of conformationally disordered alkanes, where its width increases with the chain length [5]. A broadening with increasing chain length is not observed in case of Raman jet spectra, but the signal fades into a broader background. This probably reflects the higher degree of conformational disorder for longer alkanes, in which case gauche C–C stretching vibrations may couple among each other leading to several bands scattering over a wider wavenumber range. Hairpin normal coordinates and the corresponding intensity distribution around 1080 cm^{-1} (Figs. A.17–A.19) support this conclusion. However, the overlap with the $m = 2$ C–C stretching vibration complicates the determination of the band intensity. The intensity around 1080 cm^{-1} is notably weaker in the simulated spectra because of missing contributions from multi gauche conformers.

The bands at 1111–1115 and 1144 cm^{-1} have been associated with hairpin conformers due to their appearance at $n > 18$. Examination of the simulated spectra, which involve hairpin weighting factors as deduced from the low-frequency spectra, support a hairpin assignment. A minor wavenumber deviation to scaled harmonic wavenumbers (factor 0.99) between +5 and +10 cm^{-1} (0.5–1 %) is compensated by a correct reproduction of the intensity evolution with increasing chain length and a correct prediction of the wavenumber development, namely a small blue-shift of the ≈1110 cm^{-1} band with increasing chain length and no significant chain length dependence for the 1144 cm^{-1} band. Corresponding hairpin normal coordinates of both vibrations involve substantial in-phase C–C stretching in the arms happening at the same time, while an almost Raman inactive vibration, predicted at wavenumbers in between the Raman active vibrations, is characterized by C–C stretching which is in-phase for each arm but of opposite phase considering both arms. The phase relations are clearly seen in Fig. 4.26. Assuming a two-fold sign-change of the phase at the ggtgg kink, analogous to the switch in single gauche conformers discussed above, the hairpin vibrations can be related to the all-trans m =0–2 C–C stretching vibrations. This assignment explains the wavenumber dependence on the chain length for the ≈1110 cm^{-1} hairpin band. Comparing the wavenumbers of the all-trans and hairpin $m = 0$ and $m = 2$ bands (n =19–21) yields a uniform shift of +7 cm^{-1} for $m = 0$ and of +21 cm^{-1} for $m = 2$, see Table 4.9 (p. 97). This blue-shift might stem from methylene twisting displacement involved in the hairpin vibrations, which is located partly in the kink where it is hindered by the congested surrounding.

The question remains whether these bands are specific for hairpin conformers or could have a common multi gauche component. It is not possible to exclude a small multi gauche contribution, but the $m = 0$ C–C stretching band at 1138 cm^{-1} is not seen to broaden with increasing chain length (and thus higher conformational disorder) nor does a broad background emerge at its position. Thus, if in-phase multi gauche conformer vibrations are shifted from the all-trans and single gauche $m = 0$ vibrations and overlap to the 1144 cm^{-1} band, their shift would need to be very uniform (the 1144 cm^{-1} band is quite sharp, judging from the $n = 21$ spectrum) which seems rather unlikely. For the $m = 2$ vibrations, a uniform shift would be likewise implausible, and a hairpin assignment is thus not challenged.

The bands located near 900 cm^{-1} are associated with C–C stretching (and methyl C–H bending) vibrations localized near the –CH_3 end groups, see Fig. 4.27. The position of bands in this wavenumber range is known to be a probe of the conformational sequences close to the chain ends [24, 77] with bands near 870 cm^{-1} for –tg, 850 cm^{-1} for –gt, and 890 cm^{-1} for –tt end sequences. In terms of the simulated spectra from this work, the –tg and –gt end sequences correspond to gauche 1 and 2 conformers, respectively, which indeed show Raman active vibrations close to 870 and 850 cm^{-1}, respectively. Single gauche conformers with the gauche bond closer to the chain center possess vibrations closer to the unperturbed –tt band at ≈890 cm^{-1}. All these spectral indicators appear in the jet spectra, together with an unspecific diffuse background. The by far most intense band is the –tt band found at 892–893 cm^{-1}, indicating low conformational disorder close to chain ends. More disordered alkanes in the liquid phase at room temperature give rise to a much more congested Raman spectrum around 900 cm^{-1} [5, 83].

The simulated spectra suggest that the low-frequency shoulder of the –tt band showing up at eicosane and being clearly visible at heneicosane (887 cm^{-1}) stems from the hairpin conformer. This agrees well with work from Zerbi and Gussoni who concluded from normal mode calculations ($C_{38}H_{78}$) and Raman spectroscopy

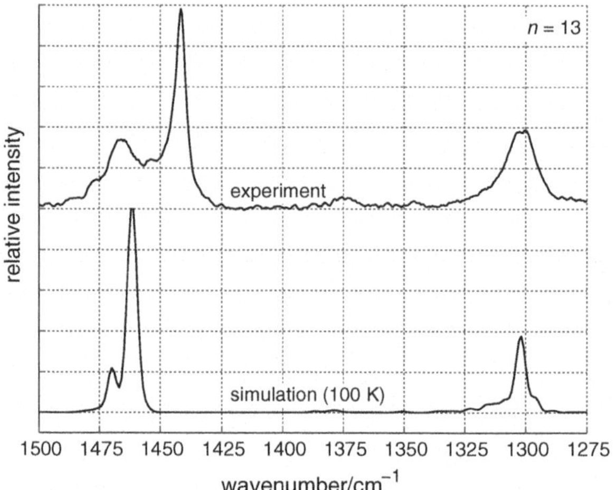

Fig. 4.28 Comparison of a jet-cooled Raman spectrum of tridecane isolated in a He expansion in the CH bending region to a B3LYP-D3/6-311++G** simulation. *Measurement conditions* $\vartheta_s = 35\,°C$, $p_0 = 0.5$bar, $p_b = 0.9$ mbar, $d_n = 1$ mm, exposure 6× 600 s. *Calculation* effective temperature used in single gauche to all-trans weighting = 100 K, wavenumber scaling factor = 0.99

of cyclic $C_{36}H_{68}$ in the solid phase that a localized ggtgg specific vibration occurs at 891 cm^{-1} [37]. The B3LYP calculations from this work confirm a partial localization to the kink, but for the much shorter chains addressed here, a strong coupling to the end group C–C stretching coordinates is found. An exemplary normal coordinate is shown in Fig. 4.27 (bottom left). In comparison with the onset of hairpin specific bands in the low-frequency region and close to the in-phase C–C stretching vibration, the delayed visibility of the 887 cm^{-1} band seems to contradict a hairpin assignment. However, this contradiction is lifted when the B3LYP calculations are examined more closely. The corresponding Raman intense hairpin vibration is found to shift to lower wavenumbers when going from $n = 13$–16 ($\approx$920 cm^{-1}) to $n = 19$–21 (901–896 cm^{-1}), with a sharp drop at $n = 17, 18$ (915,906 cm^{-1}).[13] Consequently, the associated Raman band passes through the intense –tt band when hairpin conformers start to emerge in the jet-expansion, and its absence in case of nonadecane is explained by an unresolved overlap.

The spectral region from 1400 to 1500 cm^{-1}, which was not investigated systematically, is characteristic for methylene C–H bending vibrations (scissoring) and methyl C–H bending vibrations [29]. Figure 4.28 shows a spectrum of tridecane which was recorded from a helium expansion together with a B3LYP-D3/6-311++G** simulation. The simulation reproduces the band shape qualitatively but falls short in predicting correct wavenumbers. The correct prediction of a band with two maxima might in fact be just coincidental. The methylene bending band is known to be shaped by Fermi resonance with overtones from methylene rocking vibrations (fundamentals at $\approx$720 cm^{-1}) [32] which explains the shortcomings of the harmonic calculations.

[13] Unscaled harmonic wavenumbers.

Table 4.9 Assignment of experimental C–C stretching and CH_2 deformation vibrations in cm^{-1}

Chain length									
13	14	16	17	18	19	20	21	Int.	Assignment[a]
847	847	847	846	846	848	846	846	vw–w	methyl νC–C –gt
875	870	876	871	—	878	875	—	vw	methyl νC–C –tg
—	—	—	—	—	—	—	887	w,sh	methyl νC–C –tt/–ggtgg–
893	893	892	892	892	893	892	892	s	methyl νC–C –tt
1066	1067	1067	1066	1067	1067	1066	1066	m–s	methylene νC–C $m = n - 2$
1082	1082	1082	1082	1082	1082	1081	1081	w–m	methylene νC–C –g–
—	—	—	—	1087	1090	1092	1094	w(sh)	methylene νC–C $m = 2$
—	—	—	—	—	1111	1113	1115	vw	methylene νC–C $m = 2$ (hairpin)
1109	1113	1118	1119	1121	1123	1123	1124	vw–w	methylene νC–C $m = 1$ (gauche)
1139	1140	1139	1138	1138	1138	1137	1137	s	methylene νC–C $m = 0$
—	—	—	—	—	1145	1144	1144	vw–w,sh	methylene νC–C $m = 0$ (hairpin)
1168	1169	1168	1168	1167	1168	1167	1167	w	in-phase ρCH_2
1303	1302	1300	1300	1298	1300	1298	1299	m–s,b	in-phase τCH_2

[a] ν = stretching, ρ = rocking, τ = twisting

Relative intensities (Int.) are abbreviated: b = broad, v = very, w = weak, m = medium, s = strong, sh = shoulder. C–C stretching vibrations are characterized by m-values (collective vibrations) or conformational sequences (localized vibrations), see text

[a] ν = stretching, ρ = rocking, τ = twisting

According to the simulation, the broader band at higher wavenumber (1466 cm^{-1}) should increase at high hairpin abundance against the sharper band (1442 cm^{-1}), but remain unspecific otherwise. This spectral region might be a target for further chain-folding investigations [84]. The discussed assignments are summarized in Table 4.9.

4.4.3 C–H Stretching Region

It is convenient to study alkyl chains in the C–H stretching region, because the C–H valence stretching vibrations exhibit high Raman scattering intensity and are well accessible. By analyzing C–H stretching vibrations of alkyl chains it is possible to draw conclusions regarding the molecular environment [72, 85, 86] and conformational order [80, 87]—a promising basis to study the self-solvation of gaseous n-alkanes. Raman jet spectra of the C–H stretching manifold, which occurs in the wavenumber interval 2850–2980 cm^{-1}, are depicted in Fig. 4.29. As this survey shows, the spectral changes with increasing chain length are even smaller than in the C–C stretching region. Essential features and developments are as follows. Four sharp peaks, positioned at $\approx$2860, 2870, 2890, and 2975 cm^{-1}, and the overall band shape are common to all chain lengths, whereas distinctions are apparent in complexity and band width. For n = 13–17, the spectra are almost identical, but at higher chain length two broad bands grow at $\approx$2908 and $\approx$2925 cm^{-1}, peak widths increase notably, and the peak at 2860 cm^{-1} grows a low-frequency shoulder. The onset of these spectral changes falls in the chain length interval where hairpin

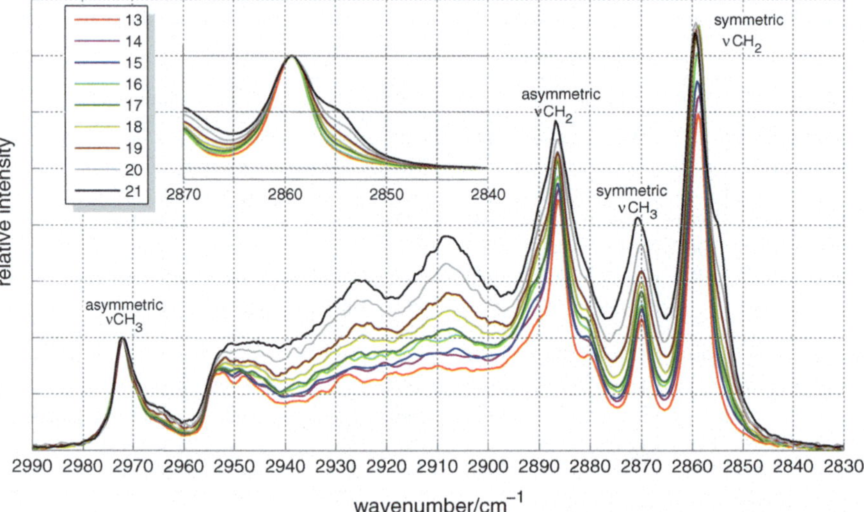

Fig. 4.29 Averaged Raman jet spectra of alkanes in He expansions in the C–H stretching region. The inlay shows a magnification of the symmetric methylene C–H stretching band, scaled on the peak intensity and shifted by not more than 1 cm^{-1} to let wavenumbers coincide for better clarity (n = 14, 15 are similar to n = 13, 16 and are omitted). The spectra are scaled to the peak intensity of the asymmetric methyl stretching band at ≈2970 cm^{-1} and Savitzky-Golay filtered (7 pt.). *Measurement conditions* exposure 28 × 90 s (n = 13, 17, 19), (60 × 90 + 72 × 60) s (n = 14), (92 × 90 + 32 × 60) s (n = 15), 24 × 90 s (n = 16, 20), 28 × 180 s (n = 18), 6 × 180 s (n = 21), otherwise analogous to low-frequency He expansion measurements. Tetradecane and pentadecane were measured more often to test reproducibility, see Sect. 4.5. Reference II—Reproduced by permission of The Royal Society of Chemistry

conformers were shown to contribute to Raman spectra at lower frequency and must thus be reviewed as possible hairpin evidence. The assignment is discussed using simulations, shown in Fig. 4.30, only in a subsidiary form, because the harmonically approximated calculations cannot grasp the complex Fermi resonance patterns characteristic for alkyl C–H stretching vibrations [71–73] (see below). The assignment relies mainly on comparison to literature spectra of liquid and solid alkanes.

Comparison to literature spectra reveals that signals of alkanes in condensed phases lie between 10 and 15 cm^{-1} lower than the jet spectra counterparts. To verify a physical origin of this shift and rule out a significant calibration error of Raman spectra, FTIR gas phase (1 mbar) and FTIR-ATR liquid phase spectra at room temperature were recorded for n = 7–10 (Fig. A.20, Appendix A.9). Measurements of longer alkanes were prevented due to insufficient volatility. FTIR vapor phase peak positions are quite close to Raman jet peak positions, so that a significant calibration error can in fact be ruled out (Fig. A.21). Beside validating the calibration of Raman spectra, the FTIR spectra were found to be a valuable link between C–H stretching spectra of the alkanes in different phases. Aided by the FTIR spectra and literature assignments [72, 80, 88, 89], the sharp peaks of the C–H stretching manifold are assigned from low to high wavenumber to the symmetric in-phase (m = 0) methylene stretching

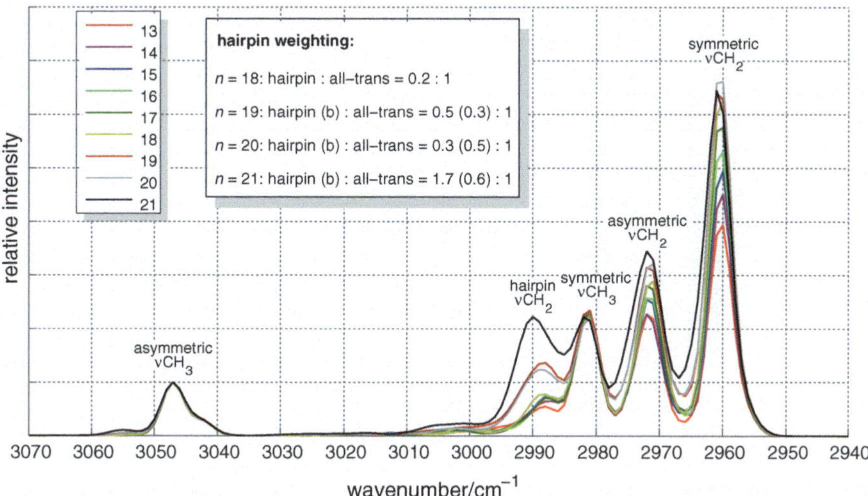

Fig. 4.30 Simulated Raman spectra from harmonic B3LYP-D3/6-311++G** calculations, see low-frequency spectra and simulations in Sect. 4.4.1. The spectra are scaled on the asymmetric CH_3 stretching vibration band, analogous to the experimental spectra. Effective temperature used in single gauche to all-trans weighting = 100 K, hairpin weighting as deduced from low frequency spectra, wavenumber scaling factor = 0.99

vibration ($\approx$2860 cm^{-1}), the symmetric methyl stretching vibration ($\approx$2870 cm^{-1}), the asymmetric in-phase ($m = 0$) methylene stretching vibration ($\approx$2890 cm^{-1}), and the nearly degenerate [88] asymmetric methyl stretching vibrations ($\approx$2970 cm^{-1}). The m-value description will be dropped at this point, because $m \neq 0$ methylene vibrations were not observed.

The symmetric methylene stretching vibration is known to take part in distinct Fermi interaction with first overtones of methylene bending vibrations ($\approx 2 \times 1440$ cm^{-1} = 2880 cm^{-1}) giving rise to an asymmetric band with a high-frequency tail running from 2850 up to 2950 cm^{-1} [71, 72, 90]. The diffuse background between 2900 and 2950 cm^{-1} in the Raman jet spectra matches the reported band shape quite well and is thus assigned to overtone or combination levels intensified by Fermi resonance with the symmetric methylene stretching vibration. The broad bands observed at $\approx$2908 and $\approx$2925 cm^{-1} for longer alkanes lie just on top of the Fermi resonance band and might be other manifestations of Fermi interaction. Their distinction from the underlying band and a possible hairpin origin is suggested when the corresponding intensities are plotted against the chain length. In Fig 4.31, the peak intensity of the 2908 and 2925 cm^{-1} band relative to the peak intensity of the asymmetric methyl stretching band (which should be approximately constant in the considered chain length interval) is plotted against the chain length. The increasing uncertainty due to stronger noise in the spectra of longer alkanes does not mask the fact that there is a slightly faster increase of intensity with chain length at a certain point. The precision of the data is not sufficient to tell whether this point is $n = 18$ or

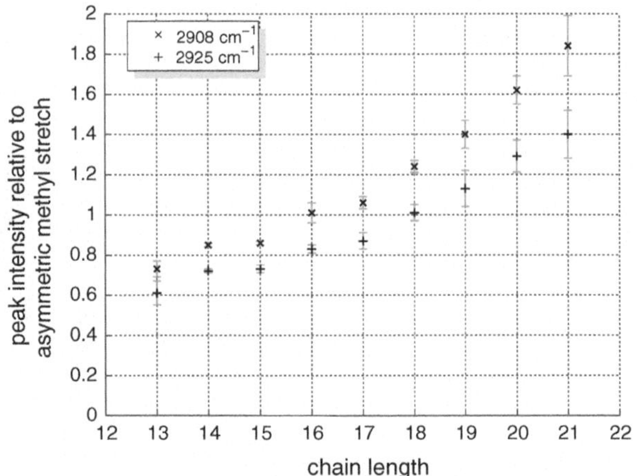

Fig. 4.31 Peak intensity ratios from the 2908 and 2925 cm^{-1} band to the asymmetric methyl stretch band (≈2970 cm^{-1}). Error bars are estimated from baseline noise and noise covering the 2908 and 2925 cm^{-1} bands. Because of the higher width of the corresponding bands, 3 data points closest to 2908 and 2925 cm^{-1} were averaged to determine the intensities at these positions

19, but the observation is certainly compatible with the previous finding that hairpin features occur in Raman jet spectra at $n = 18$. The general increase of intensity in the Fermi resonance area relative to the methyl vibration is rationalized by the fact that the intensity of the Fermi resonance band is coupled to the intensity of the symmetric methylene stretching vibration whose intensity must increase with increasing chain length compared to the methyl vibration. This treatment shows also that the peak intensity ratio of the 2908 to the 2925 cm^{-1} band stays quite constant over the given chain length interval, indicating a common origin. The harmonic B3LYP-D3/6-311++G** simulations are compatible with a hairpin assignment in so far that the main hairpin C–H stretching contribution is predicted to be positioned in the center of the C–H stretching manifold (Fig. 4.30). The associated vibration involves symmetric CH_2 stretching rather localized to the ggtgg kink. Based on these findings, the two broad bands are assigned to the hairpin conformer, but due to complex Fermi interaction and weakness of the intensity effect, this assignment must be considered tentative.

The inlay of Fig. 4.29 shows the evolution of the symmetric methylene stretching band in more detail, scaled on its peak intensity. The band starts with a symmetric shape at $n = 13$ (red) and broadens with increasing chain length. At $n > 18$, the formerly symmetric broadening proceeds notably faster and a low-frequency shoulder separates from the main band. When the spectra are examined scaled on the asymmetric methyl stretching vibration band at 2970 cm^{-1}, it is seen that the 2860 cm^{-1} band stops growing at $n = 17$, shortly before its shoulder appears (see also Fig. 4.32). These developments cannot be followed by the harmonic calculations which do not predict a shoulder in case of high hairpin abundance, nor a gradual broadening (the simulation correctly predicts the symmetric methylene stretching

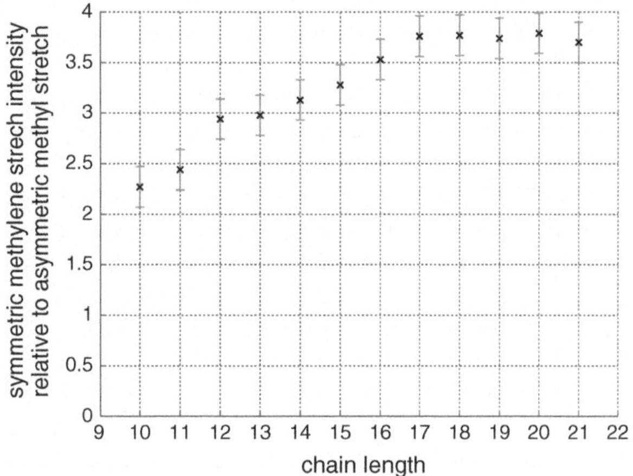

Fig. 4.32 Peak intensity of the symmetric methylene stretching vibration band at $\approx$2860 cm^{-1} relative to the peak intensity of the asymmetric methyl stretching vibration band at $\approx$2970 cm^{-1} plotted against the chain length

band to stop growing, though). The experimentally observed gradual broadening might thus be attributed to multi gauche conformers, but temperature dependent measurements reported in the literature indicate that conformational disorder does not increase the width of the symmetric methylene vibration band significantly [91].

A common multi gauche assignment to the shoulder occurring at higher chain length is again unsatisfactory, because it would imply an unlikely uniform shift. A hairpin assignment is supported by the fact that the shoulder is shifted towards the wavenumber of the symmetric methylene stretching vibration of condensed alkanes ($\approx$2850 cm^{-1}). After all, when folding into the hairpin structure, alkanes "self-solvate", and neighboring side arms are exposed to "quasi-intermolecular" interactions, so that a more liquid-like spectrum is comprehensible. The liquid-like character is what sets the hairpin conformer qualitatively apart from multi gauche conformers that have no chain segments neighboring each other. However, this character must also be appropriate for clusters, and a cluster assignment cannot be excluded.

More interesting observations are made when the widths of the symmetric methyl vibration band at 2870 cm^{-1} and the asymmetric methylene vibration band at 2886 cm^{-1} are considered. The symmetric methyl vibration band is suited for a quantitative analysis, because it keeps a symmetric shape over the whole chain length interval and can be approached by a local Gaussian curve fit. For the asymmetric methylene vibration band, a meaningful full width at half maximum (FWHM) is not defined, as the band's complex structure at short chain length merges at higher chain length. The tridecane spectrum reveals that there are actually three bands overlapping in this region: a weak one at 2880 cm^{-1}, the intense and initially sharp peak at 2886 cm^{-1}, and a broader high-frequency shoulder at $\approx$2890 cm^{-1}. Clearly, the bands broaden when the chain length increases, but whether this broadening proceeds faster at or after $n = 18$ cannot be determined unambiguously. The noise

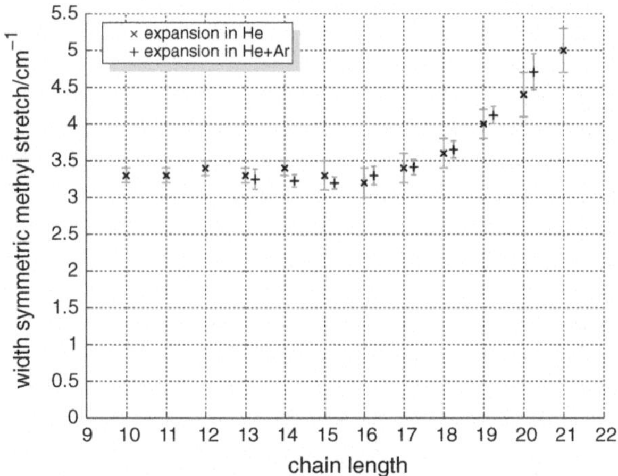

Fig. 4.33 FWHM of the symmetric methyl stretching vibration band at ≈2870 cm^{-1} plotted against the chain length. Decane to dodecane were included to further verify the constant width of this band at short chain length (He expansions). Results from measurements in He + 6 % Ar expansions are laterally shifted for better visibility. The data was acquired by locally fitting the sum of a single Gaussian curve and a linear baseline. The increasing error bars at higher chain length (provided by Matlab [92] for a 95 % confidence interval) reflect a slightly increasing deviation from the band shape to a Gaussian profile

of the spectra is too large to enable a robust curve fitting of the overlapping bands. Other than the band of the symmetric methylene stretching vibration, the asymmetric methylene stretching vibration band is known to broaden substantially when an ordered methylene system melts [32, p. 174], so that general conformational disorder with increasing chain length is a sensible explanation. Further explanations for the broadening of this band are available [72, 91], but rather restricted to the condensed phase, which would once more point at a connection to the hairpin conformer or clusters.

The situation is better for the symmetric methyl stretching vibration band at 2870 cm^{-1} which keeps a constant width over a wide chain length range. The result of fitting this band with a single Gaussian curve is depicted in Fig. 4.33 and shows that the FWHM stays constant up to $n = 16$ or 17 and increases afterwards, in line with the onset of hairpin formation deduced from the low-frequency spectra. It is reasonable that this vibration is not affected by conformational disorder, unless the methyl groups are in close contact to the each other or parts of the methylene chain. As a last comment on the C–H stretching region, the asymmetric methyl stretching vibration band shall be addressed very briefly. It is characterized by an asymmetric shape with a rather sharp peak at 2972–2973 cm^{-1} and a low-frequency tail. This tail gains intensity relative to the sharp peak, starting at $n = 17$ or $n = 18$, so that this effect might be connected to the onset of hairpin formation as well. The effect is quite small but worth mentioning. The shape of the band is reproduced by the simulation, but an increase of the intensity of the wing at high hairpin abundance is not predicted.

4.4.4 Summary: Spectral Indicators of the Hairpin Conformation

Evidence of the hairpin conformer could be successfully identified in all examined spectral regions. The low-frequency spectra were found to exhibit numerous prominent Raman bands indicative of hairpin longitudinal arm stretching vibrations as peaks of medium width covering the broad and unspecific D-LAM band at $\approx$200 cm^{-1}, with wavenumbers being largely compatible to B3LYP/6-311++G** predictions. Spectra at higher wavenumbers are less conformer specific, but hairpin specific bands were found in the region of C–C stretching vibrations nonetheless at 887, 1110–1115, and 1144 cm^{-1}, which are a mode localized to the kink and the $m = 0$ and $m = 2$ in-phase C–C stretching vibrations, respectively. The two latter are observable because they are notably shifted from their all-trans and single gauche counterparts. They occur along with the low-frequency hairpin evidence. The changes in the C–H stretching region are subtle but can be correlated partly with hairpin features at lower wavenumber when they are plotted against the chain length. By this means, broad bands growing at the center of the C–H stretching manifold (2908, 2925 cm^{-1}) could be identified as possible hairpin features. The width of the symmetric methyl vibration band at 2870 cm^{-1} increases rapidly at the onset of hairpin formation, which suggests a connection. Finally, the symmetric methylene stretching vibration band grows a low-frequency shoulder at high hairpin abundance.

From the spectral evidence, one can conclude with confidence that the hairpin conformer was indeed present in the examined supersonic jet expansions, and approach the question how it compares to the all-trans conformer concerning stability.

4.5 Critical Folding Chain Length

This section addresses the key question of this thesis: at which chain length do unbranched alkanes prefer a folded hairpin conformation over an extended all-trans structure? Since the focus of this work is rather on the experiment, an estimate of the critical folding chain length (n_c) based on Raman jet spectra will be discussed first, followed by a comparison to quantum chemical energy predictions [13, 14, 93–95]. When discussing the experimental estimate, special attention has to be given to the jet-expansion technique, which enables the preparation of low-energy alkane conformers but bears the disadvantage of producing non-equilibrium conformer distributions. The non-equilibrium conditions will become apparent when temperatures are estimated referring to Raman bands of different conformers.

4.5.1 Estimation from Raman Jet Spectra

The first experimental approach to localize the critical folding chain length, pursued at an early stage of this work, made use of the all-trans accordion vibration band and the conformationally rather independent CH-stretching band as a point of reference. Setting the integrated intensities of these bands in relation allows to estimate the overall all-trans abundance in the expansion when compared to quantum chemically calculated scattering cross-sections. Monitoring the development of the *accordion/CH ratio* with increasing chain length was expected to show that the all-trans conformer abundance would gradually decrease, due to growing competition with entropically favored gauche conformers. Furthermore, it was anticipated that the passing of the critical chain length would be accompanied by an additional drop of the all-trans abundance, when hairpin conformers would be formed at the expense of all-trans conformers.

The benefit of this indirect approach is that it is the least demanding regarding sensitivity—only strong Raman bands need to be characterized. This advantage was crucial before efforts were successful to further suppress stray light and to optimize the usage of the heatable nozzle, which eventually enabled the detection of hairpin conformers directly. However, measuring the accordion/CH ratio also holds several drawbacks which stem largely from the spectral gap between the accordion vibration and CH-stretching vibrations. None of the monochromator/grating combinations available for the curry-jet allow to measure the low-frequency accordion vibration band and the high-frequency CH-stretching band simultaneously, so that separate measurements are required. Great care has to be taken to avoid drifts of the optical alignment or fluctuation in the alkane concentration during these measurements. Effects from fluctuations in the concentration can be attenuated by measuring the two spectral regions in alternation, but drifts of the optical alignment can only be attenuated by doing so if they affect the two spectral regions equally. It turned out that the detection optics of the curry-jet are not sufficiently achromatic to guarantee the latter, causing the accordion/CH measurements to suffer from poor reproducibility.

Because of the limited success of the accordion/CH ratio measurements, the following discussion of the results will be kept rather short. Further discussion and details on underlying calculations can be found in Appendix A.6. In Fig. 4.34, the resulting ratios are plotted against the chain length. Each data point is an average of several alternating measurements, usually six spectra of the accordion vibration band (each $\sim 4 \times 300$ s) and seven spectra of the CH-stretching band (each $\sim 4 \times 90$ s). The apparent scattering of the data points demonstrates the weak reproducibility of the results and prevents the localization of a drop indicative of hairpin conformers competing with the all-trans conformer. Nevertheless, the gradual decrease of the entropically more and more handicapped all-trans conformer as well as the better cooling ability of carrier gas mixtures including heavier additives is well seen. The prediction of the accordion/CH ratio of all-trans alkanes by means of quantum chemistry allows to estimate the all-trans abundance from the experimental accordion/CH ratio, if one assumes that the overall intensity of the CH-stretching band does not depend much on the conformation.[14] The accordion/CH ratio is about 0.1 on the B3LYP-D3/6-311++G** level, which translates the experimental accordion/CH ratio to the

[14] The weak dependence of the CH-stretching band on the carrier gas supports this assumption, see Fig. 3.10 in Sect. 3.2.

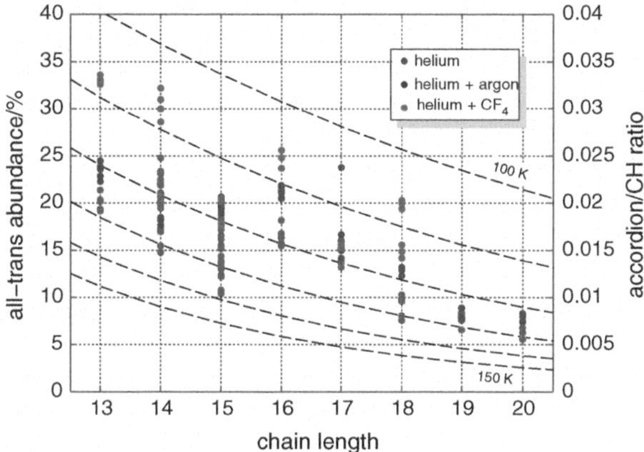

Fig. 4.34 Intensity ratio of the accordion vibration band to the CH-stretching band (accordion/CH ratio) from Raman jet measurements in different carrier gas mixtures plotted against the chain length. Calculated accordion/CH scattering cross-section ratios (B3LYP-D3/6-311++G**, Turbomole v6.4 [44]) for the $n = 13$–20 all-trans conformers at $T = 100$ K amount to ≈ 0.1, which yields the semi-empirical all-trans abundance given on the left y-axis. Calculated scattering cross-sections are polarization-weighted according to the polarization-dependent sensitivity of the curry-jet spectrograph, see Sect. 2.1. *Dashed lines* indicate the all-trans abundance according to the conformational partition function discussed in Sect. 4.1 (Eq. 4.1) at temperatures between 100–150 K (average gauche penalty $= 2.5$ kJ mol^{-1})

all-trans fraction given on the left y-axis in Fig. 4.34. This may be compared to the model partition function introduced in Sect. 4.1 (Eq. 4.1) which yields all-trans fractions indicated by dashed lines for different temperatures. According to this comparison, mean accordion/CH ratios correspond to effective conformational temperatures $T \approx 120$ K (He, He + Ar) and $T \approx 110$ K (He + CF_4), when an average gauche penalty of 2.5 kJ mol^{-1} is employed. Using lower and upper bounds of 2.0 and 3.0 kJ mol^{-1} extends these temperatures to the ranges 100 K $< T <$ 140 K (He, He + Ar) and 90 K $< T <$ 130 K (He + CF_4).

An alternative to measuring the accordion vibration band intensity relative to the CH-stretching band intensity is to measure it relative to neighboring non-all-trans signals, which eliminates problems due to chromatic aberrations and was tried subsequent to the accordion/CH measurements. In this case, however, a uniform point of reference is missing, which is why not the particular intensity ratio itself, but its temperature dependence was probed. To this end, alkanes in the size range $n = 16$–20 were measured in helium expansions and helium expansions doped with CF_4 to further lower the temperature. In all cases, the accordion vibration band gained intensity by $\approx (30 \pm 10)$ % compared to its spectral surrounding when CF_4 was added to the expansion, that is, the all-trans abundance went up by roughly the same amount. Again, no footprint of hairpin conformers was found, which in this case should have been a smaller increase of all-trans abundance upon CF_4 addition when the hairpin conformer starts to compete. This experiment and the former accordion/CH ratio measurements lead to the conclusion that the hairpin conformer does in fact not

Table 4.10 (1) Effective conformational temperatures (*T*/K) of jet-isolated alkanes with chain length n, estimated from the experimental conformer abundance $\left(\frac{N_i}{N_j}\right)$ and relative B3LYP-D3/6-311++G** energies (ΔE^0/kJ mol^{-1}, ZPVE corrected) using Eq. 4.8. (2) Hairpin to all-trans abundance ratios, absolute and relative to an isoenergetic crossing-point

(1) *Conformational temperatures*					
	n	N_i/N_j	ΔE^0		
Overall all-trans	13–20	0.1–0.3	—	$140 > T > 90$	
(From accordion/CH ratio)					
Single gauche to all-trans[a]	13–21	0.05–0.3	2.1–2.9	$110 > T > 80$	
Hairpin b to hairpin	19	0.6	2.6	$T > 300$	
	20	1.7	0.8	$T > 180$	
	21	0.35	2.0	$T > 170$	
(2) *Experimental hairpin to all-trans conformer ratios*					
	$n \leq 17$	$n = 18$	19	20	21
Hairpin (absolute)	< 0.1	0.2	0.5	0.3	1.7
(Relative to isoenergetic crossing-point) (%)	<5	10	13	15	43
Hairpin b (absolute)	—	—	0.3	0.5	0.6
(Relative to isoenergetic crossing-point) (%)	—	—	8	13	15

[a] See Table 4.4

compete with the all-trans conformer *in supersonic expansions* of alkanes in the size range up to $n = 20$. Does this mean that the critical chain length is not yet reached?

Fortunately, the spectra from the gain measurements were of sufficient quality to allow the observation of signals compatible with quantum chemically predicted hairpin vibrations. Quantum chemical calculations were initially restricted to all-trans and hairpin conformers and later extended to single gauche conformers to establish the validity of harmonic B3LYP frequencies, and to rule out alternative assignments challenging the hairpin assignments. Moreover, the simulations allowed to estimate conformer fractions relative to each other and to monitor the fraction of hairpin conformers with increasing chain length in particular. During the discussion of the assignment, the conformer fractions were addressed occasionally and shall be reviewed here in more detail.

The accuracy of the simulation was addressed in Sect. 4.3.4. It should be recalled that conformer fractions determined from the simulations do not reach quantitative quality, which is why their discussion is kept rather on a qualitative level. Conformational fractions and corresponding temperatures are summarized in Table 4.10. The listed effective single gauche temperature (entry *single gauche to all-trans*) and temperature based on the accordion/CH ratio were previously discussed and are close to 100 K. The temperatures determined from accordion/CH ratios tend to be somewhat higher than single gauche temperatures, which could stem from higher gauche conformers frozen in the rapidly cooled expansion, but the temperature difference is not large enough to conclude this with certainty. When temperatures are derived from hairpin abundance ratios, however, (entry *hairpin b to hairpin*) significantly higher values are found. The lower bounds given in the table already account for an uncertainty of 30 % in the conformer ratio. Calculating with the listed values yields even higher temperatures closer to the nozzle temperature. This suggests that hairpin conformers freeze in an early, warmer stage of the expansion, and are not significantly populated once the expansion reaches effective temperatures close to values that describe the all-trans and single gauche population. This behavior makes sense and can be explained by the very steep temperature gradient of the expansion. Isomerization from a random gauche to a hairpin conformation certainly needs many collisions, since the generation of the specific ggtgg sequence, correctly positioned in the chain and of uniform helicity, is statistically not very likely. Relaxation rather leads to depopulation of conformers which could isomerize to hairpin structures with relatively few trans-gauche isomerizations. Therefore, most often conformational isomerization will cease before the hairpin conformation can be "found". It is thus even conceivable that a large portion of hairpin conformers observed in jet-expansions is frozen from the gas phase distribution at the nozzle temperature, and merely pre-folded conformers (for example, a hairpin conformer with one gauche conformation near the end of the chain) efficiently relax to completely aligned hairpin conformers. The situation is quite the opposite for the all-trans conformer, which can be formed extensively from single and double gauche conformers which are abundant at low temperatures. As a consequence, the all-trans to hairpin conformer ratio in jet-expansions should be largely determined by kinetic factors, while thermodynamic stability plays a secondary role. This explains why measurements based merely on the abundance of the all-trans conformer (accordion/CH ratio, gain of accordion vibration) do not provide evidence for self-solvation. Hairpin conformers do in fact not compete significantly with all-trans conformers, because their formation is kinetically delayed, but this does not imply that they are less stable.

The second part of Table 4.10 summarizes hairpin/all-trans conformer ratios. Only the heneicosane hairpin conformer outweighs the all-trans conformer when absolute conformer fractions are compared. However, absolute ratios are somewhat misleading because they do not account for crucial enantiomeric degeneracy and symmetry weighting. The achiral all-trans conformer has a statistical disadvantage compared to hairpin conformers which can be realized in two enantiomeric forms. If one accounts for the symmetry number and enantiomeric degeneracy, the hairpin conformer should be twice as abundant (C_2 symmetric hairpin) or even four times as abundant as the all-trans conformer (C_1 symmetric hairpin) in a thermally equilibrated system, if all-trans and hairpin conformers are just isoenergetic. Therefore, the interesting quantity is rather the hairpin/all-trans ratio relative to the isoenergetic point of reference. Such values are listed in the table below the absolute ratios, assuming C_2 symmetry for hairpin conformers of alkanes with even-numbered chain length and C_1 symmetry otherwise (includes hairpin b conformers). Accounting for statistical weighting, none of the hairpin conformers outweigh the abundance of the all-trans conformer. However, there is a steady increase of hairpin abundance with increasing chain length starting at $n = 18$, including even less stable hairpin conformers for $n \geq 19$, which must be explained by the hairpin conformer becoming energetically feasible. The jump of hairpin abundance in case of $n = 21$ might be related to differing expansion conditions,[15] but it could also stem from pre-folded conformers, which might be abundant in the warm gas phase at this high chain length, and are able to relax to hairpin conformers efficiently.

In the light of a kinetic delay which slows down self-solvation, it is most sensible to assign the critical chain length to the onset of hairpin formation. The hairpin evidence is scarce for octadecane, but overwhelming for nonadecane. One might thus assign either $n_c = 17$ or, more conservatively, $n_c = 18$. However, octadecane benefits from folding a bit more than nonadecane where the last methyl group of the hairpin conformer does not have a direct neighbor, so that $n_c = 17$ is somewhat more likely. This assessment is in agreement with most recent quantum chemical energy predictions at high computational levels, which will be outlined among other calculations in the next section.

To narrow down the energy of hairpin conformers, one might consider abundances of single gauche conformers, which have a well known energetic disadvantage to the all-trans structure, and anticipate some results of energy calculations. Even the least stable hairpin conformer observed, the $n = 19$ hairpin b conformer, is as abundant as single gauche conformers, which sets ≈ 2.5 kJ mol^{-1} as an upper bound for its energy relative to all-trans. Quantum chemical calculations consistently predict that hairpin conformers become more stable by at least 1or 2 kJ mol^{-1} per chain segment (depending on whether the chain length switches from an even to an odd number, or *vice versa*), so that it appears very unlikely that all hairpin conformers up to $n = 21$ remain in the energy range between 0 and 2.5 kJ mol^{-1}. Furthermore, the zero-point corrected energy difference of the $n = 19$ hairpin and hairpin b conformer is 2.64 or 2.70 kJ mol^{-1} on the B3LYP- D3/6-311++G** or B3LYP-D3/def2-TZVP level, respectively. From these relative hairpin energies (which are probably more reliable than predicted energies relative to all-trans conformers) follows that the $n = 19$ hairpin conformer is a little more stable than the corresponding all-trans conformer, which supports $n_c \leq 18$.

4.5.2 Computational Predictions

The self-solvation of linear alkanes was recently addressed in several computational studies published by Goodman [13], Thomas et al. [14], and Grimme et al. [93, 94]. The most recent computational effort was published by Mata together with experimental results from this work [95] and is probably the most reliable prediction to date. The employed computational methods reach from fast force field calculations (Goodman, Thomas), over dispersion corrected DFT and MP2 calculations (Grimme) up to costly coupled cluster calculations (Mata). The obtained critical folding chain lengths are summarized in Table 4.11, together with dispersion corrected DFT results from this work and the experimental estimate.

[15] That is, a high amount of air in the expansion, sucked in by the brass saturator. If so, this would invite for further measurements with helium mixed with a substantial amount of nitrogen.

Table 4.11 Critical chain length according to different computational methods and Raman jet spectra Reference II—Reproduced by permission of The Royal Society of Chemistry

Method	Reference	Critical chain length
MM2	[13]	17
MM3	[13]	24
AMBER	[13]	25
OPLS-AA	[14]	21 (15–17)[a]
B97-D[b]	This work	12
B3LYP-D3[c]	This work	15
B3LYP-D3[d]	This work	16
MP2+CC	[95]	17 ± 1
Raman jet	**This work**	**17–18**

[a]estimated accounting for known deficiencies of OPLS-AA [14]
[b]6-311++G(d,p) basis set, Gaussian 09 [96]
[c]6-311++G** basis set, Turbomole v6.4 [44]
[d]def2-TZVP basis set, Turbomole v6.4 [44]

The critical chain length provided by Grimme et al. [93, 94] is not directly comparable to this work or the other computational studies because it deals with a different hairpin geometry, folded by a $x^{\pm}g^{\mp}x^{\pm}$ sequence (xgx in the following). In this work, the xgx hairpin was compared to the ggtgg hairpin on the B3LYP-D3/def2-TZVP level for $n = 20$ and found to be less stable by 13.0 kJ mol^{-1}. According to calculated Raman spectra, it would be difficult to distinguish the two conformers by their spectroscopic signatures, which is why the presence of the xgx hairpin in supersonic expansions cannot be excluded judging from the experimental spectra alone. However, the xgx hairpin is almost certainly less stable than a corresponding ggtgg hairpin, not only because its chain ends are less well aligned, but also because of the unfavorable congestion of the xgx sequence. Coupled cluster conformer energies of smaller alkanes [18] predict that two $g^{\pm}g^{\pm}$ sequences cost roughly 8 kJ mol^{-1} while a $x^{\pm}g^{\mp}x^{\pm}$ sequence costs about 20 kJ mol^{-1} (less stable by $\approx$12 kJ mol^{-1}) which agrees very well with the B3LYP-D3 energy difference of the xgx and ggtgg hairpin mentioned above (13.0 kJ mol^{-1}). With such a high energetic disadvantage, the xgx hairpin will play no role in the studied supersonic expansions, at least not in the considered size range. The B3LYP-D3 energy of the $n = 20$ xgx hairpin relative to all-trans (+5.9 kJ mol^{-1}, including ZPVE) is at variance with MP2 and DFT-D (BLYP-D, B97-D [52]) calculations published by Grimme, which consistently predict that the $n = 20$ xgx hairpin is more stable than the corresponding all-trans conformer.[16] Likely, the MP2 and DFT-D methods employed by Grimme would yield a shorter critical chain length than the value reported for the xgx hairpin ($n_c = 14$–21) [93, 94] when applied to the ggtgg hairpin. Indeed, B97-D/6-311++G(d,p) predictions, which were calculated in an early stage of this work with Gaussian 09 [96] find $n_c = 12$, clearly too short compared with the experimental estimate and the coupled cluster calculations from Mata (see below). This points to an overestimation

[16] Based on a linear extrapolation of the published energies for $n = 14$, 22, and 30.

of the dispersion attraction by the 2006 DFT-D version [52] and/or an underestimation of the xgx conformer energy. One should emphasize that the goal of Refs. [93] and [94] was not to pinpoint the exact switching point, but to emphasize its existence and the importance of dispersion corrections to locate it.

Goodman [13] calculated the energy of ggtgg hairpin conformers relative to all-trans conformers using several popular force fields (MM2, MM3, AMBER) as well as semi-empirical methods (PM3, AM1, MNDO). The latter give critical chain lengths far from the experimental turning point (or none at all, MNDO) and are not listed in Table 4.11. Unsuccessfully seeking non-extended alkane conformers that would surpass the stability of the all-trans conformer by random conformational searches, the ggtgg folding motive was eventually imposed by Goodman, which demonstrates the low probability of finding the hairpin conformer by chance and supports the assumption that many collisions are necessary to form hairpin conformers from random gauche conformers in supersonic expansions. The force field predictions are $n_c = 17$ for MM2, 24 for MM3, and 25 for AMBER. With AMBER, Goodman modeled solvent effects as well, yielding the reasonable results that a polar solvent would promote self-solvation (less favorable interactions with the solvent are minimized) while an apolar solvent would dramatically postpone it, in line with the estimated late onset of chain folding in crystalline alkanes [97] (interactions with neighboring molecules outweigh the self-solvation energy), which could still be kinetically controlled.

OPLS-AA force field calculations from Thomas et al. [14] fall in between the MM2, MM3, and AMBER predictions ($n_c = 21$). However, Thomas et al. noted that the gauche energy yielded from OPLS-AA calculations on $g^{\pm}g^{\pm}$ (gg) *n*-pentane—an important test case for the energy penalty of the ggtgg sequence—is somewhat overestimated, and that invoking more accurate *ab inito* gauche energies [2] suggests a critical chain length in the range 15–17. It is instructive to do the same comparison for MM2, MM3, and AMBER. All these force fields calculate a gg pentane energy of $\approx$6.7 kJ mol^{-1},[17] somewhat less then OPLS-AA, which yields 7.9 kJ mol^{-1}, but still significantly larger than results from precise coupled cluster calculations (3.8–4.2 kJ mol^{-1} [2, 4, 18]) and an experimental estimate (3.9 kJ mol^{-1} [19]). Therefore, the force fields overestimate the ggtgg energy penalty (and/or underestimate the "positive pentane effect" [2], see Sect. 4.1) which explains the general overestimation of the critical chain length compared with the experiment and most recent coupled cluster prediction [95]. The match of the MM2 result despite the overestimated gauche penalty suggests an overestimation of van-der-Waals attraction as well, which was indeed corrected in the later version MM3 [98]. The comparison of MM2 and MM3 in this aspect emphasizes the sensitivity and usefulness of the alkane folding-benchmark.

The first-principles approach from Mata [95] is based on MP2 and coupled cluster theory (and thus abbreviated MP2+CC in the following) employing density fitting and local correlation (DF-LMP2 [99], DF-LCCSD(T) [100]) to reduce the computational expense, which would otherwise prevent the high-level description of large alkanes.

[17] MM2: Ref. [21], MM3: Ref. [45], AMBER energy calculated with Gaussian 09 [96].

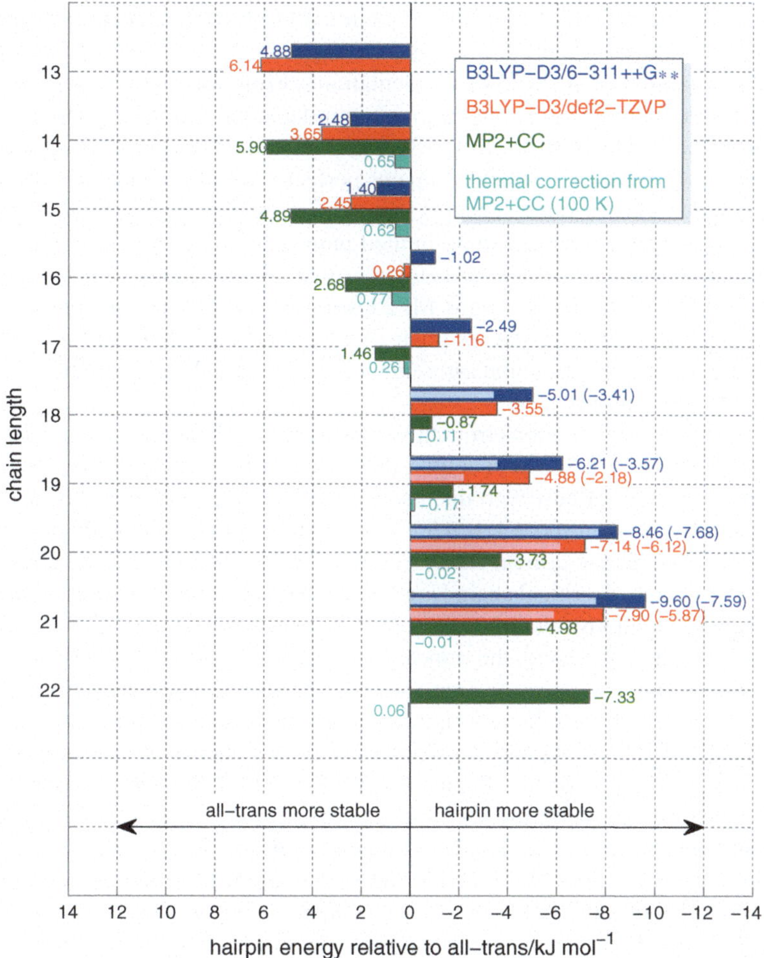

Fig. 4.35 Energy of the hairpin conformer relative to the all-trans conformer. *Blue*, *red* and *green bars* comprise electronic and zero-point vibrational energy from B3LYP-D3/6-311++G**, B3LYP-D3/def2-TZVP, and a composite MP2/coupled cluster calculation (MP2+CC) [95]. Energies from hairpin b conformers are drawn with *light colors* and are given in parentheses. *Blue-green bars* are thermal corrections at 100 K from the MP2+CC calculations. Reference II—Reproduced by permission of The Royal Society of Chemistry

In this approach, electronic energies of DF-LMP2/cc-pVTZ structures are refined treating higher-order electron correlation at the DF-LCCSD(T) level to correct at least energetically for the known overestimation of dispersion attraction by MP2 methods [101–103]. The resulting hairpin energies from this approach are given in Fig. 4.35, including zero-point corrections derived from scaled harmonic frequency calculations on the DF-LMP2/cc-pVTZ level. Thermal corrections on the same level of theory for $T = 100$ K are depicted separately. These could not be calculated in case of the B3LYP-D3 calculations from this work, due to faulty predictions of vibrations with very low-frequencies (see Sect. 4.3), but the DF-LMP2 calculations suggest a very small thermal contribution which can be neglected without making a significant error. The reliability of the MP2 + CC calculation scheme was tested using *n*-pentane conformers yielding energies with a maximum deviation[18] of 0.23 kJ mol^{-1} from recent coupled cluster values [2, 4, 18]. The MP2 + CC energy of the gg pentane conformer is 4.01kJ mol^{-1} which compares perfectly with the available experimental and coupled cluster values ($\approx$4.0 kJ mol^{-1}, see above). The accuracy from the MP2+CC approach is claimed to be limited mainly by the uncertainty in the zero-point vibrational energy, which leads to an estimated uncertainty of ± 1 in n_c [95]. However, structural deviations at the MP2 level used for the chain geometry cannot be ruled out. The final MP2 + CC estimate of the critical chain length is $n_c = 17 \pm 1$, which validates the experimental assessment that the critical chain length is reached at 17 or 18 carbon atoms.

The B3LYP-D3 calculations from this work were mainly intended to aid the spectral assignment (Sect. 4.3), but of course, they also provide a critical chain length and allow an assessment of how well the dispersion correction performs. The approach used to simulate Raman spectra, B3LYP-D3 in combination with the Pople basis set 6-311++G**, yields the critical chain length $n_c = 15$, which is already quite close to the best estimate. A critical chain length still a little closer to the experimental and coupled cluster values results when the more complete basis set def2-TZVP is chosen ($n_c = 16$). On the other hand, the truncation of the basis set to def-SV(P) lowers the critical chain length to $n_c = 14$ (not shown). This is a consequence from using a small basis set which introduces a significant intramolecular basis set superposition error (BSSE) [104] and thereby artificially lowers the energy of the compact hairpin conformer. The less BSSE affected B3LYP-D3/def2-TZVP calculations seem to very satisfactorily describe the torsional and dispersion energy of folded alkanes. This could stem from favorable error compensation, like in the case for the MM2 force field, but other than MM2, B3LYP-D3 calculations also perform well regarding the energies of small alkanes (the gg pentane energy is 4.45 kJ mol^{-1} on the B3LYP-D3/def2-TZVP level, see also hexane conformer energies in Appendix A.1) which underlines the validity of this approach.

Calculations where the energy difference between hairpin and all-trans is available (this excludes the publications from Goodman [13] and Thomas et al. [14]) show that the trend with chain length is fundamentally linear with a slight alternation superimposed. This will not hold for hairpin conformers of rather short alkanes, but it is true in the important size range and might be exploited to extrapolate the critical chain length without calculating energies of very large alkanes in the future. A plot demonstrating the linear trend of the hairpin energy is shown in Appendix A.9, Fig. A.22.

It was anticipated in the previous section that B3LYP-D3 and MP2+CC calculations yield very similar results when it comes to the relative energy evolution of hairpin conformers, which can be extracted from the numbers given in Fig. 4.35. From the MP2+CC calculations follows that increasing the chain length by one

[18] Applies to the $x^{\pm}g^{\mp}$ pentane conformer with the highest energy relative to tt pentane.

segment from an even to an odd number lowers the hairpin energy relative to all-trans on average by (1.1 ± 0.2) kJ mol^{-1}, while increasing the chain length from an odd to an even number lowers the energy by (2.2 ± 0.2) kJ mol^{-1}. The B3LYP-D3 results from both used basis sets are virtually identical: (1.2 ± 0.3) kJ mol^{-1} (even → odd) and (2.3 ± 0.2) kJ mol^{-1} (odd → even), respectively. From these robust energy increments, one would expect the hairpin abundance observed in the experiment to increase more steeply with increasing chain length. The fact that this is not the case is another indication of the kinetic delay discussed above.

References

1. G. Tasi, F. Mizukami, I. Pálinkó, J. Csontos, W. Győrffy, P. Nair, K. Maeda, M. Toba, S.-I. Niwa, Y. Kiyozumi, I. Kiricsi, Enumeration of the conformers of unbranched aliphatic alkanes. J. Phys. Chem. A **102**, 7698–7703 (1998)
2. J.B. Klauda, B.R. Brooks, A.D. MacKerell, R.M. Venable, R.W. Pastor, An ab initio study on the torsional surface of alkanes and its effect on molecular simulations of alkanes and a DPPC bilayer. J. Phys. Chem. B **109**, 5300–5311 (2005)
3. G. Tasi, B. Nagy, G. Matisz, T.S. Tasi, Similarity analysis of the conformational potential energy surface of *n*-pentane. Computat. Theor. Chem. **963**, 378–383 (2011)
4. J.M.L. Martin, What can we learn about dispersion from the conformer surface of *n*-pentane? J. Phys. Chem. A **117**, 3118–3132 (2013)
5. D.A. Cates, H.L. Strauss, R.G. Snyder, Vibrational Modes of liquid *n*-alkanes: simulated isotropic Raman spectra and band progressions for C_5H_{12}-$C_{20}H_{42}$ and $C_{16}D_{34}$. J. Phys. Chem. **98**, 4482–4488 (1994)
6. T. N. Wassermann, Umgebungseinflüsse auf die C-C- und C-O-Torsionsdynamik in Molekülen und Molekülaggregaten: Schwingungsspektroskopie bei tiefen Temperaturen, Ph.D. Thesis, Georg-August-Universität Göttingen, 2009
7. H. Gotō, E. Ōsawa, M. Yamato, How many conformers are there for small *n*-alkanes? Consequences of asymmetric deformation in GG' segment. Tetrahedron **49**, 387–396 (1993)
8. T.N. Wassermann, J. Thelemann, P. Zielke, M.A. Suhm, The stiffness of a fully stretched polyethylene chain: a Raman jet spectroscopy extrapolation. J. Chem. Phys. **131**, 161108 (2009)
9. E. Funck, Symmetrie und Anzahl der Rotationsisomere von unverzweigten Polymerketten. Zeitschrift für Elektrochemie **62**, 901–905 (1958)
10. A. Abe, R.L. Jernigan, P.J. Flory, Conformational energies of *n*-alkanes and the random configuration of higher homologs including polymethylene. J. Am. Chem. Soc. **88**, 631–639 (1966)
11. G. Tasi, F. Mizukami, Quantum algebraic-combinatoric study of the conformational properties of *n*-alkanes I. J. Math. Chem. **25**, 55–64 (1999)
12. G. Tasi, F. Mizukami, J. Csontos, W. Győrffy, I. Pálinkó, Quantum algebraic-combinatoric study of the conformational properties of *n*-alkanes. J. Math. Chem. **27**, 191–199 (2000)
13. J.M. Goodman, What Is the longest unbranched alkane with a linear global minimum conformation? J. Chem. Inf. Comput. Sci. **37**, 876–878 (1997)
14. L.L. Thomas, T.J. Christakis, W.L. Jorgensen, Conformation of alkanes in the gas phase and pure liquides. J. Phys. Chem. B **110**, 21198–21204 (2006)
15. J. Šebek, L. Pele, E.O. Potma, R. Benny Gerber, Raman spectra of long chain hydrocarbons: anharmonic calculations, experiment and implications for imaging of biomembranes. Phys. Chem. Chem. Phys. **13**, 12724–12733 (2011)

16. R.S. Judson, E.P. Jaeger, A.M. Treasurywala, M.L. Peterson, Conformational searching methods for small molecules. II. Genetic algorithm approach. J. Comput. Chemi. **14**, 1407–1414 (1993)
17. S.M. Long, T.T. Tran, P. Adams, P. Darwen, M.L. Smythe, Conformational searching using a population-based incremental learning algorithm. J. Comput. Chemi. **32**, 1541–1549 (2011)
18. D. Gruzman, A. Karton, J.M.L. Martin, Performance of ab initio and density functional methods for conformational equilibria of C_nH_{2n+2} alkane isomers ($n = 4$–8). J. Phys. Chem. A **113**, 11974–11983 (2009)
19. R.M. Balabin, Enthalpy difference between conformations of normal alkanes: Raman spectroscopy study of n-pentane and n-butane. J. Phys. Chem. A **113**, 1012–1019 (2009)
20. S. Bocklitz, Conformational analysis of n-alkanes in cryosolutions, M.Sc. Thesis, Georg-August-Universität Göttingen, Universiteit Antwerpen, 2013
21. S. Tsuzuki, L. Schäfer, H. Gotō, E.D. Jemmis, H. Hosoya, K. Siam, K. Tanabe, E. Ōsawa, Investigation of intramolecular interactions in n-alkanes. Cooperative energy increments associated with GG and GTG' sequences. J. Am. Chem. Soc. **113**, 4665–4671 (1991)
22. E. Koglin, R.J. Meier, An ab-initio study of the relative stability of the ggg and the gtg conformer in hexane. Chem. Phys. Lett. **312**, 284–290 (1999)
23. D. Šatkovskienė, Theoretical justification of empirical additivity schemes for conformational energies. Int. J. Quantum Chem. **91**, 5–12 (2003)
24. R.G. Snyder, Chain conformation from the direct calculation of the Raman spectra of the liquid n-alkanes C12–C20. J. Chem. Soc. Faraday Trans. **88**, 1823–1833 (1992)
25. L. Brambilla, G. Zerbi, Local Order in Liquid n-Alkanes: Evidence from Raman Spectroscopic Study. Macromolecules **38**, 3327–3333 (2005)
26. W.A. Herrebout, B.J. van der Veken, A. Wang, J.R. Durig, Enthalpy Difference between conformers of n-butane and the potential function governing conformational interchange. J. Phys. Chem. **99**, 578–585 (1995)
27. M. Lipton, W.C. Still, The multiple minimum problem in molecular modeling. Tree searching internal coordinate conformational space. J. Comput. Chem. **9**, 343–355 (1988)
28. R. Zbinden, *Infrared Spectroscopy of High Polymers* (Academic Press, New York, 1964)
29. R.G. Snyder, J.H. Schachtschneider, Vibrational analysis of the n-paraffins - I: Assignments of infrared bands in the spectra of C_3H_8 through n-$C_{19}H_{40}$. Spectrochim. Acta **19**, 85–116 (1963)
30. C. Kittel, *Einführung in die Festkörperphysik*, 14th edn. (Oldenbourg-Verlag, München, 2006)
31. M. Tasumi, T. Shimanouchi, T. Miyazawa, Normal vibrations and force constants of polymethylene chain. J. Mol. Spectrosc. **9**, 261–287 (1962)
32. G. Zerbi (ed.), *Modern Polymer Spectroscopy* (Wiley-VCH, Weinheim, 1999)
33. L. Piseri, G. Zerbi, A generalization of GF method to crystal vibrations. J. Mol. Spectrosc. **26**, 254–261 (1968)
34. T. Shimanouchi, Local and overall vibrations of polymer chains. Pure Appl. Chem. **36**, 93–108 (1973)
35. R.G. Snyder, Vibrational study of the chain conformation of the liquid n-paraffins and molten polyethylene. J. Chem. Phys. **47**, 1316–1360 (1967)
36. G. Zerbi, L. Piseri, F. Cabassi, Vibrational spectrum of chain molecules with conformational disorder: polyethylene. Mol. Phys. **22**, 241–256 (1971)
37. G. Zerbi, M. Gussoni, Defect modes for (200), GGTGG, tight fold re-entry in polyethylene single crystals. Polymer **21**, 1129–1134 (1980)
38. R.G. Snyder, The structure of chain molecules in the liquid state: low-frequency Raman spectra of n-alkanes and perfluoro-n-alkanes. J. Chem. Phys. **76**, 3921–3927 (1982)
39. R.G. Snyder, S.L. Wunder, Long-range conformational structure and low-frequency isotropic Raman spectra of some highly disordered chain molecules. Macromolecules **19**, 496–498 (1986)
40. R.F. Schaufele, Chain shortening in polymethylene liquids. J. Chem. Phys. **49**, 4168–4175 (1968)

41. S.-I. Mizushima, T. Simanouti, Raman frequencies of *n*-paraffin molecules. J. Am. Chem. Soc. **71**, 1320–1324 (1949)
42. R.F. Schaufele, T. Shimanouchi, Longitudinal acoustical vibrations of finite polymethylene chains. J. Chem. Phys. **47**, 3605–3610 (1967)
43. R.G. Snyder, H.L. Strauss, R. Alamo, L. Mandelkern, Chain-length dependence of interlayer interaction in crystalline *n*-alkanes from Raman longitudinal acoustic mode measurements. J. Chem. Phys. **100**, 5422–5431 (1994)
44. TURBOMOLE v6.4 2012, a development of University of Karlsruhe and Forschungszentrum Karlsruhe GmbH, 1989–2007, TURBOMOLE GmbH, since 2007; available from http://www.turbomole.com
45. N.L. Allinger, Y.H. Yuh, J.H. Lii, Molecular mechanics. The MM3 force field for hydrocarbons. 1. J. Am. Chem. Soc. **111**, 8551–8566 (1989)
46. A.D. Becke, Density-functional thermochemistry. III. The role of exact exchange. J. Chem. Phys. **98**, 5648–5652 (1993)
47. C. Lee, W. Yang, R.G. Parr, Development of the Colle-Salvetti correlation-energy formula into a functional of the electron density. Phys. Rev. B **37**, 785–789 (1988)
48. S.H. Vosko, L. Wilk, M. Nusair, Accurate spin-dependent electron liquid correlation energies for local spin density calculations: a critical analysis. Can. J. Phys. **58**, 1200–1211 (1980)
49. P.J. Stephens, F.J. Devlin, C.F. Chabalowski, M.J. Frisch, Ab initio calculation of vibrational absorption and circular dichroism spectra using density functional force fields. J. Phys. Chem. **98**, 11623–11627 (1994)
50. M.D. Wodrich, C. Corminboeuf, P. von Ragué Schleyer, Systematic errors in computed alkane energies using B3LYP and other popular DFT functionals. Org. Lett. **8**, 3631–3634 (2006)
51. S. Grimme, Accurate description of van der Waals complexes by density functional theory including empirical corrections. J. Comput. Chem. **25**, 1463–1473 (2004)
52. S. Grimme, Semiempirical GGA-type density functional constructed with a long-range dispersion correction. J. Comput. Chem. **27**, 1787–1799 (2006)
53. S. Grimme, J. Antony, S. Ehrlich, H. Krieg, A consistent and accurate ab initio parametrization of density functional dispersion correction (DFT-D) for the 94 elements H-Pu. J. Chem. Phys. **132**, 154104 (2010)
54. R. Krishnan, J.S. Binkley, R. Seeger, J.A. Pople, Self-consistent molecular orbital methods. XX. A basis set for correlated wave functions. J. Chem. Phys. **72**, 650–654 (1980)
55. T. Clark, J. Chandrasekhar, G.W. Spitznagel, P. von Ragué Schleyer, Efficient diffuse function-augmented basis sets for anion calculations. III. The 3–21+G basis set for first-row elements, Li-F. J. Comput. Chem. **4**, 294–301 (1983)
56. A. Schäfer, H. Horn, R. Ahlrichs, Fully optimized contracted Gaussian basis sets for atoms Li to Kr. J. Chem. Phys. **97**, 2571–2577 (1992)
57. K. Eichkorn, O. Treutler, H. Öhm, M. Häser, R. Ahlrichs, Auxiliary basis sets to approximate Coulomb potentials. Chem. Phys. Lett. **242**, 652–660 (1995)
58. F. Weigend, A fully direct RI-HF algorithm: Implementation, optimised auxiliary basis sets, demonstration of accuracy and efficiency. Phys. Chem. Chem. Phys. **4**, 4285–4291 (2002)
59. F. Weigend, M. Häser, H. Patzelt, R. Ahlrichs, RI-MP2: optimized auxiliary basis sets and demonstration of efficiency. Chem. Phys. Lett. **294**, 143–152 (1998)
60. K. Eichkorn, F. Weigend, O. Treutler, R. Ahlrichs, Auxiliary basis sets for main row atoms and transition metals and their use to approximate Coulomb potentials. Theoret. Chem. Acc. **97**, 119–124 (1997)
61. A.P. Scott, L. Radom, Harmonic vibrational frequencies: an evaluation of Hartree-Fock, Møller-Plesset, quadratic configuration interaction, density functional theory, and semiempirical scale factors. J. Phys. Chem. **100**, 16502–16513 (1996)
62. M.W. Wong, Vibrational frequency prediction using density functional theory. Chem. Phys. Lett. **256**, 391–399 (1996)
63. C.J. Cramer, *Essentials of Computational Chemistry: Theories and Models*, 2nd edn. (John Wiley & Sons Ltd, Chichester, 2004)

64. Y. Zhao, D.G. Truhlar, Density functionals with broad applicability in chemistry. Acc. Chem. Res. **41**, 157–167 (2008)
65. C. Van Caillie, R.D. Amos, Raman intensities using time dependent density functional theory. Phys. Chem. Chem. Phys. **2**, 2123–2129 (2000)
66. J. Neugebauer, M. Reiher, B.A. Hess, Coupled-cluster Raman intensities: assessment and comparison with multiconfiguration and density functional methods. J. Chem. Phys. **117**, 8623–8633 (2002)
67. E.E. Zvereva, A.R. Shagidullin, S.A. Katsyuba, Ab initio and DFT predictions of infrared intensities and Raman activities. J. Phys. Chem. A **115**, 63–69 (2011)
68. M.D. Halls, H.B. Schlegel, Comparison study of the prediction of Raman intensities using electronic structure methods. J. Chem. Phys. **111**, 8819–8824 (1999)
69. R.G. Snyder, Y. Kim, Conformation and low-frequency isotropic Raman spectra of the liquid *n*-alkanes C_4-C_9. J. Phys. Chem. **95**, 602–610 (1991)
70. K.S. Smirnov, D. Bougeard, Quantum-chemical derivation of electro-optical parameters for alkanes. J. Raman Spectrosc. **37**, 100–107 (2006)
71. R.G. Snyder, J.R. Scherer, Band structure in the C-H stretching region of the Raman spectrum of the extended polymethylene chain: Influence of Fermi resonance. J. Chem. Phys. **71**, 3221–3228 (1979)
72. R.G. Snyder, H.L. Strauss, C.A. Elliger, Carbon-hydrogen stretching modes and the structure of *n*-alkyl chains. 1. Long, disordered chains. J. Phys. Chem. **86**, 5145–5150 (1982)
73. N.A. Atamas, A.M. Yaremko, T. Seeger, A. Leipertz, A. Bienko, Z. Latajka, H. Ratajczak, A.J. Barnes, A study of the Raman spectra of alkanes in the Fermi-resonance region. J. Mol. Struct. **708**, 189–195 (2004)
74. H.-P. Grossmann, H. Bölstler, Longitudinal acoustic modes in cycloalkanes. Polym. Bull. **5**, 175–177 (1981)
75. H.-P. Grossmann, Investigation of conformational transitions in cycloalkanes, especially $(CH_2)_{22}$. Polym. Bull. **5**, 137–144 (1981)
76. H.F. Kay, B.A. Newman, The crystal and molecular structure of cyclotetratriacontane $[CH_2]_{34}$. Acta Crystallogr. B **24**, 615–624 (1968)
77. M. Maroncelli, S.P. Qi, H.L. Strauss, R.G. Snyder, Nonplanar conformers and the phase behavior of solid *n*-alkanes. J. Am. Chem. Soc. **104**, 6237–6247 (1982)
78. S. Wolf, C. Schmid, P.C. Hägele, Vibrational analysis of the tight (110) fold in polyethylene. Polymer **31**, 1222–1227 (1990)
79. R.J. Meier, A. Csiszár, E. Klumpp, On the interpretation of the 1100 cm^{-1} Raman band in phospholipids and other alkyl-containing molecular entities. J. Phys. Chem. B **110**, 5842–5844 (2006)
80. C.J. Orendorff, M.W. Ducey Jr, J.E. Pemberton, Quantitative correlation of Raman spectral indicators in determining conformational order in alkyl chains. J. Phys. Chem. A **106**, 6991–6998 (2002)
81. G.R. Strobl, W. Hagedorn, Raman spectroscopic method for determining the crystallinity of polyethylene. J. Polym. Sci. Polym. Phys. Ed. **16**, 1181–1193 (1978)
82. Y. Morino, K. Kuchitsu, A note on the classification of normal vibrations of molecules. J. Chem. Phys. **20**, 1809–1810 (1952)
83. G. Zerbi, R. Magni, M. Gussoni, K.H. Moritz, A. Bigotto, S. Dirlikov, Molecular mechanics for phase transition and melting of *n*-alkanes: a spectroscopic study of molecular mobility of solid *n*-nonadecane. J. Chem. Phys. **75**, 3175–3194 (1981)
84. K.-S. Lee, G. Wegner, S.L. Hsu, Vibrational spectroscopic studies of linear and cyclic alkanes C_nH_{2n+2}, C_nH_{2n} with $24 \leq n \leq 288$: Chain folding, chain packing and conformations. Polymer **28**, 889–896 (1987)
85. B.P. Gaber, W.L. Peticolas, On the quantitative interpretation of biomembrane structure by Raman spectroscopy, Biochimica et Biophysica Acta **465**, 260–274 (1977)
86. P.B. Miranda, V. Pflumio, H. Saijo, Y.R. Shen, Chain-chain interaction between surfactant monolayers and alkanes or alcohols at solid/liquid interfaces. J. Am. Chem. Soc. **120**, 12092–12099 (1998)

87. R.G. Snyder, On Raman evidence for conformational order in liquid n-alkanes. J. Chem. Phys. **76**, 3342–3343 (1982)
88. R.A. MacPhail, R.G. Snyder, H.L. Strauss, The motional collapse of the methyl C-H stretching vibration bands. J. Chem. Phys. **77**, 1118–1137 (1982)
89. R.A. MacPhail, H.L. Strauss, R.G. Snyder, C.A. Elliger, Carbon-hydrogen stretching modes and the structure of n-alkyl chains. 2. Long, all-trans chains. J. Phys. Chem. **88**, 334–341 (1984)
90. S. Abbate, G. Zerbi, S.L. Wunder, Fermi resonances and vibrational spectra of crystalline and amorphous polyethylene chains. J. Phys. Chem. **86**, 3140–3149 (1982)
91. S.L. Wunder, M.I. Bell, G. Zerbi, Band broadening of CH_2 vibrations in the Raman spectra of polymethylene chains. J. Chem. Phys. **85**, 3827–3839 (1986)
92. MATLAB, version 7.12.0 (R2011a), The MathWorks Inc., Natick (2011)
93. S. Grimme, J. Antony, T. Schwabe, C. Mück-Lichtenfeld, Density functional theory with dispersion corrections for supramolecular structures, aggregates, and complexes of (bio)organic molecules. Org. Biomol. Chem. **5**, 741–758 (2007)
94. T. Schwabe, S. Grimme, Double-hybrid density functionals with long-range dispersion corrections: higher accuracy and extended applicability. Phys. Chem. Chem. Phys. **9**, 3397–3406 (2007)
95. N.O.B. Lüttschwager, T.N. Wassermann, R.A. Mata, M.A. Suhm, The last globally stable extended alkane. Angew. Chemie Int. Ed. **52**, 463–466 (2013)
96. M.J. Frisch, G.W. Trucks, H.B. Schlegel, G.E. Scuseria, M.A. Robb, J.R. Cheeseman, G. Scalmani, V. Barone, B. Mennucci, G.A. Petersson, H. Nakatsuji, M. Caricato, X. Li, H.P. Hratchian, A.F. Izmaylov, J. Bloino, G. Zheng, J.L. Sonnenberg, M. Hada, M. Ehara, K. Toyota, R. Fukuda, J. Hasegawa, M. Ishida, T. Nakajima, Y. Honda, O. Kitao, H. Nakai, T. Vreven, J.A. Montgomery, Jr., J.E. Peralta, F. Ogliaro, M. Bearpark, J.J. Heyd, E. Brothers, K.N. Kudin, V.N. Staroverov, R. Kobayashi, J. Normand, K. Raghavachari, A. Rendell, J.C. Burant, S.S. Iyengar, J. Tomasi, M. Cossi, N. Rega, J.M. Millam, M. Klene, J.E. Knox, J.B. Cross, V. Bakken, C. Adamo, J. Jaramillo, R. Gomperts, R.E. Stratmann, O. Yazyev, A.J. Austin, R. Cammi, C. Pomelli, J.W. Ochterski, R.L. Martin, K. Morokuma, V.G. Zakrzewski, G.A. Voth, P. Salvador, J.J. Dannenberg, S. Dapprich, A.D. Daniels, O. Farkas, J.B. Foresman, J.V. Ortiz, J. Cioslowski, D.J. Fox, Gaussian 09, Revision A.02, Gaussian Inc, Wallingford CT, 2009
97. G. Ungar, J. Stejny, A. Keller, I. Bidd, M.C. Whiting, The crystallization of ultralong normal paraffins: the onset of chain folding. Science **229**, 386–389 (1985)
98. J.H. Lii, N.L. Allinger, Molecular mechanics. The MM3 force field for hydrocarbons. 3. The van der Waals' potentials and crystal data for aliphatic and aromatic hydrocarbons. J. Am. Chem. Soc. **111**, 8576–8582 (1989)
99. H.-J. Werner, F.R. Manby, P.J. Knowles, Fast linear scaling second-order Møller-Plesset perturbation theory (MP2) using local and density fitting approximations. J. Chem. Phys. **118**, 8149–8160 (2003)
100. H.-J. Werner, M. Schütz, An efficient local coupled cluster method for accurate thermochemistry of large systems. J. Chem. Phys. **135**, 144116 (2011)
101. J. Grant Hill, J.A. Platts, H.-J. Werner, Calculation of intermolecular interactions in the benzene dimer using coupled-cluster and local electron correlation methods. Phys. Chem. Chem. Phys. **8**, 4072–4078 (2006)
102. K.E. Riley, P. Hobza, Assessment of the MP2 Method, along with several basis sets, for the computation of interaction energies of biologically relevant hydrogen bonded and dispersion bound complexes. J. Phys. Chem. A **111**, 8257–8263 (2007)
103. R.A. Bachorz, F.A. Bischoff, S. Höfener, W. Klopper, P. Ottiger, R. Leist, J.A. Frey, S. Leutwyler, Scope and limitations of the SCS-MP2 method for stacking and hydrogen bonding interactions. Phys. Chem. Chem. Phys. **10**, 2758–2766 (2008)
104. R.M. Balabin, Enthalpy difference between conformations of normal alkanes: effects of basis set and chain length on intramolecular basis set superposition error. Mol. Phys. **109**, 943–953 (2011)

Chapter 5
Perfluorinated Alkanes

Associated with this work, experiments on perfluorinated alkanes were conducted in the context of a bachelor thesis [1]. The main goal of this research project was the extrapolation of the elastic modulus of polytetrafluoroethylene (PTFE)—the polymer counterpart of the investigated oligomers (C_nF_{2n+2} with $n = 6, 8–10, 12–14$), but the accumulated spectra also provide a valuable comparative case for normal alkanes, regarding energetics and structural preferences. This topic will be discussed in the present chapter, the elastic modulus is outlined in the following Chap. 6.

A striking difference between n-alkanes and their perfluorinated counterparts is the ground state structure of the latter: in the crystalline state, long PTFE chains do not assume a planar all-trans, but helical structures [2, 3]. Two energetically nearly degenerate states are relevant at ambient conditions [3, 4], with helical chains having 6 turns per 13 carbon atoms (13/6 helix) or 7 turns per 15 carbon atoms (15/7 helix) in the repeating unit. The former is also denoted form II[1] and is observed at temperatures below 19 °C, the latter, form IV, at temperatures above 19 °C. A transition to the all-trans structure (2/1 helix) can be induced at high pressures [6]. In the gas phase, short perfluorinated alkanes prefer a structure close to the 13/6 helix, with C–C–C–C torsional angles between 162° and 163° [7, 8]. Figure 5.1 shows a corresponding helical structure of perfluorotetradecane, the longest oligomer measured in the course of this work. The "source of helicity in perfluorinated n-alkanes" [9] was explained by the higher steric demand of fluorine compared to hydrogen atoms [10] (van-der-Waals radii of 1.47 and 1.10 Å, respectively [11, 12]), and electrostatic interactions [9], both causing –CF_2– units in 1,3 position to avoid each other, and thus a planar trans conformation.

[1] The conformation of form II PTFE is only close to a 13/6 helix, but the deviations are very small [5].

N.O.B. Lüttschwager, *Raman Spectroscopy of Conformational Rearrangements at Low Temperatures*, Springer Theses, DOI 10.1007/978-3-319-08566-1_5

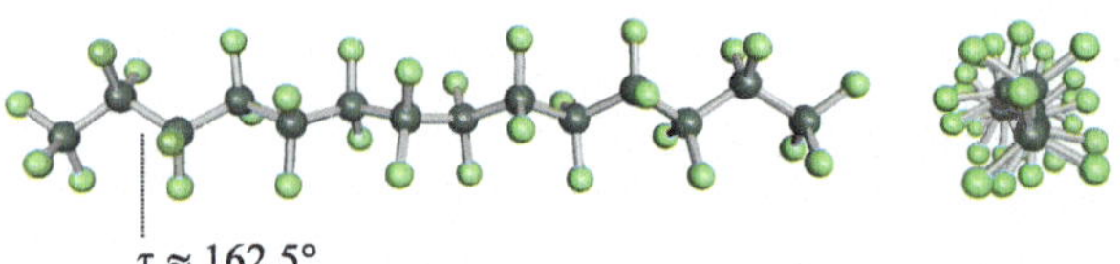

Fig. 5.1 Minimum energy structure of $C_{14}F_{30}$ optimized on the B3LYP-D3/def2-TZVP level (Turbomole v6.4 [13]). All torsional angles (τ) are twisted away from the trans conformation by $\approx$17.5°. Reproduced and adapted material from reference III

5.1 Raman Jet Spectra

The torsional isomerism of perfluorinated *n*-alkanes is more complicated compared to the hydrocarbon counterparts, because there are further stable conformations, denoted anti ($\tau \approx \pm 162°$), ortho ($\tau \approx \pm 95°$), and gauche ($\tau \approx \pm 55°$) [14–17]. Other than in case of normal alkanes, the ortho ("cross") conformation can occur independently from the adjacent conformations. From an entropic point of view, one would thus expect to observe more conformational variety in spectra of jet-isolated perfluoroalkanes. Yet, the experimental finding is quite the opposite. Raman jet spectra (Fig. 5.2) show almost no indications of conformers other than the anti-helix with $\tau \approx \pm 162°$. The measurement conditions vary somewhat from those under which normal alkanes were measured, but previous measurements of normal alkanes [18] suggest that the influence of the differing nozzle temperature and stagnation pressure is rather minor. The origin of the mono-conformational perfluoroalkane spectra must be explained otherwise.

One apparent aspect is the energy difference between anti and gauche conformers. Qualitatively, reported experimental observations underline a higher gauche penalty: Low-frequency Raman spectra of liquid dodecane and perfluorododecane show that the LAM-1 band stands out against the D-LAM band much clearer in case of the latter [19]. In general, bandwidths in low-frequency spectra of liquid perfluorinated alkanes are smaller than those in spectra of normal alkanes [20]. Quantitative estimations of the gauche energy penalty are 4.6 kJ mol^{-1} [21] and 5.1 kJ mol^{-1} [22]. Gas phase spectra recorded with the curry-jet at $\approx$300 °C [1] are too congested to allow a reliable assessment of the anti/gauche energy difference, but they show a more pronounced LAM-1 band when compared to previous gas phase measurements of normal alkanes [18] as well. A higher gauche energy is thus strongly supported by the experiment.

MP2 calculations of electronic energies tend to predict values close to the trans/gauche energy difference of normal alkanes, $\approx$2 kJ mol^{-1} [14, 23, 24], while DFT calculations yield higher values of about $\approx$4 kJ mol^{-1} [3, 16]. B3LYP energies from this work depend on the usage of Grimme's D3 dispersion correction [25]. Electronic energy differences calculated without D3 correction are close to 4 kJ mol^{-1}, but inclusion of D3 lowers the gauche energy penalty to about $\approx$2.5 kJ mol^{-1} (Appendix A.1). Thus, judging from the experiment, D3 corrects in the wrong direction in this case, contrary to normal alkanes, where inclusion of D3 yields a more

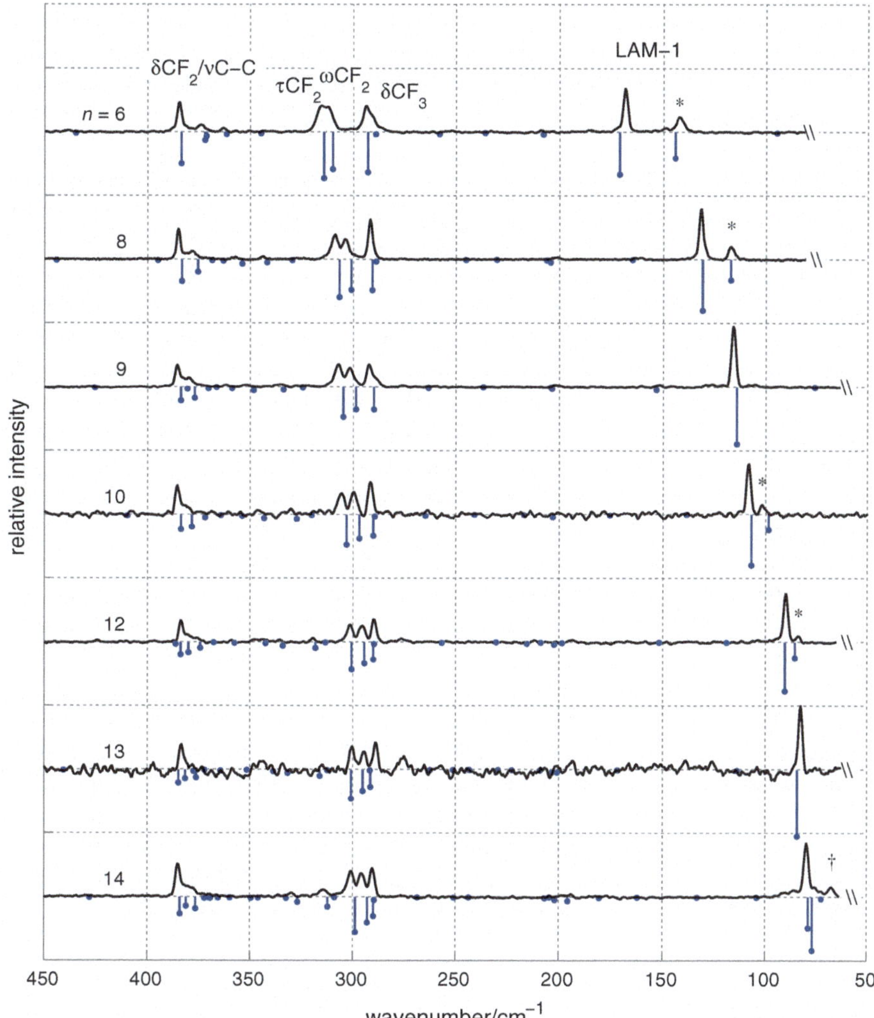

Fig. 5.2 Averaged low-frequency Raman jet spectra of perfluorinated alkanes in He expansions, set against calculated scattering cross-sections (*blue stems*). The spectra are scaled on the overall accordion vibration band intensity. Bands of modes mixing with the accordion vibration are marked with an asterisk. The $n = 14$ accordion vibration band overlaps with rotational lines from air impurities (†). Spectra are Savitzky-Golay filtered (7 pt.). *Measurement conditions* (n = 6–14) $p_0 = 0.8$ bar, $p_b \approx 2.0$ mbar, $d_n = 1$ mm; ($n = 6$) ≈ 0.5 % in He, exposure 6×150 s; ($n = 8$) ≈ 0.5 % in He, exposure 6×100 s; ($n = 9$) ≈ 0.5 % in He, exposure 6×60 s; ($n = 10$) $\vartheta_s = -7$ °C, exposure 6×60 s; ($n = 12$) $\vartheta_s = 25$ °C, exposure 6×300 s; ($n = 13$) $\vartheta_s = 25$ °C, exposure 6×600 s; ($n = 14$) $\vartheta_s = 55$ °C, exposure 12×200 s. n = 6–9 mixtures with He were prepared in a vacuum line, n = 10–13 mixtures with the glass saturator, and $n = 14$ with the heatable brass saturator. *Calculation* global minimum structure, B3LYP-D3/def2-TZVP, Turbomole v6.4 [13], $T = 100$ K. Reproduced and adapted material from reference III

accurate gauche energy. High-precision coupled cluster calculations of perfluorinated alkanes are challenging because of the large number of electrons carried by the fluorines. A recent study found a CCSD/cc-pVTZ energy difference of 3.1 kJ mol^{-1} [17] for perfluorobutane, in between non dispersion-corrected DFT results and MP2 results.

Besides arguments based on energy minima, comparing normal alkanes and perfluorinated alkanes with respect to energy barriers yields a kinetic argument for the difference in the jet conformer-population. From the torsional energy profile of perfluorinated alkanes [14, 24] one can see that the ortho energy-minimum effectively lowers the torsional energy barrier between the gauche and anti conformation ($\approx$8 kJ mol^{-1} [16, 24]). The barrier separating gauche and trans conformations in hydrocarbons is significantly higher ($\approx$12 kJ mol^{-1} [26–28]). Therefore, while freezing due to rapid cooling leads to a notable residual gauche population in supersonic expansions of normal alkanes, the conformational distribution of perfluorinated alkanes will probably freeze at lower temperatures, leading to a negligible gauche population. The nearly mono-conformational perfluoroalkane jet spectra are thus comprehensible from a kinetic and thermodynamic viewpoint.

Calculated spectra of the lowest energy 13/6 helix conformers are included in Fig. 5.2. The match of unscaled harmonic B3LYP-D3/def2-TZVP wavenumbers and experimental band positions is excellent, while calculated intensities agree qualitatively with the experiment. The comparison shows that indeed no significant contribution of conformers other than the 13/6 helix is observed (the weak signals surrounding the perfluorotetradecane LAM-1 band stem from rotational transitions of residual air). The bands are readily assigned using the quantum chemical predictions and literature spectra [20–22]: chain length dependent LAM-1 and coupling modes (the latter marked with an asterisk) are found between 70–170 cm^{-1}, chain length independent –CF_2– wagging (ωCF_2) and twisting (τCF_2) modes lie at $\approx$300 cm^{-1}, together with a C–F bending end group vibration (δCF_3). Vibrations with –CF_2– bending and C–C stretching character ($\delta CF_2/\nu$C–C) lie at $\approx$380 cm^{-1}.

5.2 Self-Solvation

In the context of this thesis, it seems appropriate to briefly discuss the matter of self-solvation. Although no indications of folding are observed in the jet spectra, the capability of the heatable nozzle was not fully exploited with perfluorotetradecane, measured at a preparation temperature of 55 °C. Judging from boiling points of normal alkanes and perfluorinated alkanes, the longest chain length accessible with the curry-jet might be about $n = 25$. Is it likely to detect hairpin-like conformers of perfluorinated alkanes at such chain lengths? This question should be discussed with regard to the gauche energy penalty, stabilizing dispersion interactions, and geometrical constraints.

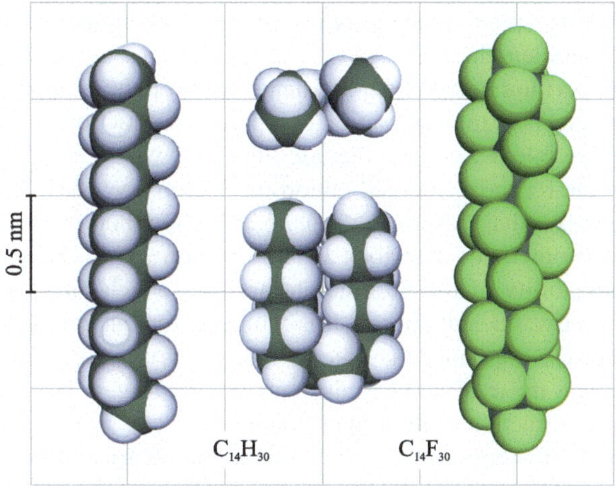

Fig. 5.3 Size comparison of all-trans (*left*) and hairpin (*center*) tetradecane to 13/6 helical perfluorotetradecane (*right*) using space filling molecule drawings (atom radius set to van-der-Waals radius, as provided by the Turbomole user interface TmoleX [29]). Structures are calculated on the B3LYP-D3/6-311++G** level, and on the B3LYP-D3/def2-TZVP level, respectively

Geometrical constraints are apparent when considering molecule drawings of alkanes and perfluorinated alkanes in correct scale, with atom sizes set to van-der-Waals radii (Fig. 5.3). This picture underlines that the ggtgg sequence is ideal to tightly pack a single normal alkane chain, providing a maximum of stabilizing dispersion interactions. The distance of hydrogen atoms of the neighboring arms is just slightly larger (≈0.25 nm) than twice their van-der-Waals radius (0.11 nm [11, 12]). A perfluorinated chain, on the other hand, demands more space, due to the larger C–F bond length ($d_{C\text{–}F} \approx 0.14$ nm versus $d_{C\text{–}H} \approx 0.11$ nm), van-der-Waals radius of fluorine (0.14 nm [11, 12]), and the helical structure (see also Fig. A.14, Appendix A.7). A "tight fold" [30] ggtgg sequence would lead to a steric clash of chain segments and is thus not transferable to perfluorinated alkanes. The ggtgg sequence by itself is probably unfavorably congested. This is indicated by calculated conformer energies of perfluoroalkanes (Appendix A.1). For normal alkanes, neighboring gauche bonds of equal sign are known to be slightly stabilized, that is, the corresponding energy is less than twice the energy of trans-separated gauche bonds (Sect. 4.1). In case of perfluorinated alkanes, the calculated conformer energies suggest the opposite: double gauche conformers with neighboring gauche bonds of equal sign have an energy higher than the sum of corresponding single gauche energies.[2]

From this discussion it is clear that the conformation of a self-solvating perfluoroalkane must be fundamentally different from the hydrocarbon analog. Since the C–C bond distance in perfluorinated and normal alkanes is almost identical, a kink that allows a larger chain separation must involve more $–CF_2–$ units than the eight

[2] See Fig. A.2, aag, aga, and agg conformers.

–CH_2– units making up an alkane ggtgg sequence. This immediately postpones a critical folding chain length.

Provided that perfluorinated chain segments are brought to appropriate contact, the question remains how their attraction compares to attraction in normal alkanes. This would involve interactions of "organic fluorine", which are a current topic of research [31] in the recently emerged "fluorous" chemistry[3] [33]. A key feature of fluorous compounds is that they are hydrophobic and lipophobic at the same time—a property which is sometimes described as "fluorous effect" [34] and often ascribed to the low polarizability of the fluorine atoms [34], which bind electrons tightly due to a high effective nuclear charge. It is tempting to conclude from this general behavior of fluorous compounds that interactions of two perfluorinated alkyl chains will be weaker than those of alkyl chains carrying hydrogen. However, compared to hydrogen, the contraction of fluorine is counterbalanced by the higher number of electrons, so that both atoms end up with comparable polarizabilities [35].

To assess how interactions of the two systems compare, one might examine their boiling points. The smallest perfluorinated alkanes with $n < 4$ have higher boiling points than normal alkanes in this size range [36], indicating a stronger interaction energy which is lost at higher chain length. This trend was rationalized by calculations of dimerization energies: Ab initio MP2 calculations on smaller dimers ($n = 1$–3) [36] found that $(CF_4)_2$ and $(C_2F_6)_2$ are more stable than hydrocarbon analogons, but the energy preference decreases quickly and switches already at $n = 3$. The authors ascribed this development to increasing dimer separation with increasing chain length because of higher steric demands. The higher stability of $(CF_4)_2$ compared with $(CH_4)_2$ was confirmed by high-precision coupled cluster calculations [37]. Investigations of longer perfluoroalkane dimers are rare due to the high computational costs. A study on (perfluoro-)hexane dimers employing force field and a semi-empirical calculations yielded contrary results regarding whether the hydrocarbon or perfluorinated dimer is more stable [35]. However, the more trustworthy calculations rather suggest, that dispersion stabilization is larger for normal alkanes, at least at a system size relevant to self-solvation.

In summary, perfluorinated alkanes cannot be folded as sharply as normal alkanes, implying a different (bigger) kink-conformation, as well as weaker stabilizing dispersion interactions. A critical folding chain length is postponed further by the higher gauche energy penalty. This leads to the conclusion that it is very unlikely to observe hairpin-like conformers of perfluorinated alkanes anywhere near the size range where normal alkanes were found to self-solvate. On the other hand, the improved conformational relaxation may assist their detection in a supersonic jet expansion, once they are energetically competitive.

[3] Fluorous: 'of, relating to, or having the characteristics of highly fluorinated saturated organic materials, molecules or molecular fragments' [32].

References

1. P. Drawe, Mechanical properties of chain molecules from spectroscopic data, B.Sc. Thesis, Georg-August-Universität Göttingen, 2010
2. C.W. Bunn, E.R. Howells, Structures of molecules and crystals of fluoro-carbons. Nature **174**, 549–551 (1954)
3. C. Quarti, A. Milani, C. Castiglioni, Ab initio calculation of the IR spectrum of PTFE: helical symmetry and defects. J. Phys. Chem. B **117**, 706–718 (2013)
4. M. D'Amore, G. Talarico, V. Barone, Periodic and high-temperature disordered conformations of polytetrafluoroethylene chains: an ab initio modeling. J. Am. Chem. Soc. **128**, 1099–1108 (2006)
5. E. Clark, The molecular conformations of polytetrafluoroethylene: forms II and IV. Polymer **40**, 4659–4665 (1999)
6. R.G. Brown, Vibrational spectra of polytetrafluoroethylene: effects of temperature and pressure. J. Chem. Phys. **40**, 2900–2908 (1964)
7. J.A. Fournier, R.K. Bohn, J.A. Montgomery Jr, M. Onda, Helical C_2 structure of perfluoropentane and the C_{2v} structure of perfluoropropane. J. Phys. Chem. A **114**, 1118–1122 (2010)
8. J.A. Fournier, C.L. Phan, R.K. Bohn, Microwave spectroscopy and characterization of the helical conformer of perfluorohexane. Arch. Org. Chem. (Part V), 5–11 (2011)
9. S.S. Jang, M. Blanco, W.A. Goddard III, G. Caldwell, R.B. Ross, The source of helicity in perfluorinated n-alkanes. Macromolecules **36**, 5331–5341 (2003)
10. M. Iwasaki, On the helical structure of polytetrafluoroethylene. J. Polym. Sci. Part A: Gen. Pap. **1**, 1099–1104 (1963)
11. M. Mantina, A.C. Chamberlin, R. Valero, C.J. Cramer, D.G. Truhlar, Consistent van der Waals radii for the whole main group. J. Phys. Chem. A **113**, 5806–5812 (2009)
12. A. Bondi, van der Waals volumes and radii. J. Phys. Chem. **68**, 441–451 (1964)
13. TURBOMOLE v6.4 2012, a development of University of Karlsruhe and Forschungszentrum Karlsruhe GmbH, 1989–2007, TURBOMOLE GmbH, since 2007; available from http://www.turbomole.com
14. G.D. Smith, R.L. Jaffe, D.Y. Yoon, Conformational characteristics of poly(tetrafluoroethylene) chains based upon ab initio electronic structure calculations on model molecules. Macromolecules **27**, 3166–3173 (1994)
15. B. Albinsson, J. Michl, Anti, ortho, and gauche conformers of perfluoro-n-butane: matrix-isolation IR spectra and calculations. J. Phys. Chem. **100**, 3418–3429 (1996)
16. U. Röthlisberger, K. Laasonen, M.L. Klein, M. Sprik, The torsional potential of perfluoro n-alkanes: a density functional study. J. Chem. Phys. **104**, 3692–3700 (1996)
17. M.R. Munrow, R. Subramanian, A.J. Minei, D. Antic, M.K. MacLeod, J. Michl, R. Crespo, M.C. Piqueras, M. Izuha, T. Ito, Y. Tatamitani, K. Yamanou, T. Ogata, S.E. Novick, Rotational spectra of gauche perfluoro-n-butane, C_4F_{10}; perfluoro-iso-butane, $(CF_3)_3CF$; and tris(trifluoromethyl)methane, $(CF_3)_3CH$. J. Mol. Spectro. **242**, 129–138 (2007)
18. T.N. Wassermann, Umgebungseinflüsse auf die C-C- und C-O-Torsionsdynamik in Molekülen und Molekülaggregaten: Schwingungsspektroskopie bei tiefen Temperaturen, Ph.D. Thesis, Georg-August-Universität Göttingen, 2009
19. R.G. Snyder, The structure of chain molecules in the liquid state: low-frequency Raman spectra of n-alkanes and perfluoro-n-alkanes. J. Chem. Phys. **76**, 3921–3927 (1982)
20. J.F. Rabolt, B. Fanconi, Longitudinal acoustic modes of polytetrafluoroethylene copolymers and oligomers. Polymer **18**, 1258–1264 (1977)
21. S.L. Hsu, N. Reynolds, S.P. Bohan, H.L. Strauss, R.G. Snyder, Structure, crystallization, and infrared spectra of amorphous perfluoro-n-alkane films prepared by vapor condensation. Macromolecules **23**, 4565–4575 (1990)
22. M. Campos-Vallette, M. Rey-Lafon, Vibrational spectra and rotational isomerism in short chain n-perfluoroalkanes. J. Mol. Struct. **101**, 23–45 (1983)
23. E.K. Watkins, W.L. Jorgensen, Perfluoroalkanes: conformational analysis and liquid-state properties from ab initio and Monte Carlo calculations. J. Phys. Chem. A **105**, 4118–4125 (2001)

24. O. Borodin, G.D. Smith, D. Bedrov, A quantum chemistry based force field for perfluoroalkanes and poly(tetrafluoroethylene). J. Phys. Chem. B **106**, 9912–9922 (2002)
25. S. Grimme, J. Antony, S. Ehrlich, H. Krieg, A consistent and accurate ab initio parametrization of density functional dispersion correction (DFT-D) for the 94 elements H-Pu. J. Chem. Phys. **132**, 154104 (2010)
26. W.A. Herrebout, B.J. van der Veken, A. Wang, J.R. Durig, Enthalpy difference between conformers of n-butane and the potential function governing conformational interchange. J. Phys. Chem. **99**, 578–585 (1995)
27. N.L. Allinger, J.T. Fermann, W.D. Allen, H.F. Schaefer III, The torsional conformations of butane: definitive energetics from ab initio methods. J. Chem. Phys. **106**, 5143–5150 (1997)
28. Y. Mo, A critical analysis on the rotation barriers in butane. J. Org. Chem. **75**, 2733–2736 (2010)
29. C. Steffen, K. Thomas, U. Huniar, A. Hellweg, O. Rubner, A. Schroer, TmoleX—a graphical user interface for TURBOMOLE. J. Comput. Chem. **31**, 2967–2970 (2010)
30. G. Zerbi, M. Gussoni, Defect modes for (200), GGTGG, tight fold re-entry in polyethylene single crystals. Polymer **21**, 1129–1134 (1980)
31. R. Berger, G. Resnati, P. Metrangolo, E. Weber, J. Hulliger, Organic fluorine compounds: a great opportunity for enhanced materials properties. Chem. Soc. Rev. **40**, 3496–3508 (2011)
32. J. Gladysz, D.P. Curran, Fluorous chemistry: from biphasic catalysis to a parallel chemical universe and beyond. Tetrahedron **58**, 3823–3825 (2002)
33. I.T. Horváth (ed.), *Fluorous Chemistry* (Springer, Heidelberg, 2011)
34. M. Cametti, B. Crousse, P. Metrangolo, R. Milani, G. Resnati, The fluorous effect in biomolecular applications. Chem. Soc. Rev. **41**, 31–42 (2012)
35. J.D. Dunitz, A. Gavezzotti, W.B. Schweizer, Molecular shape and intermolecular liaison: hydrocarbons and fluorocarbons. Helv. Chim. Acta **86**, 4073–4092 (2003)
36. S. Tsuzuki, T. Uchimaru, M. Mikami, S. Urata, Magnitude and orientation dependence of intermolecular interaction of perfluoropropane dimer studied by high-level ab initio calculations: comparison with propane dimer. J. Chem. Phys. **121**, 9917–9924 (2004)
37. M.J. Biller, S. Mecozzi, A high level computational study of the CH_4/CF_4 dimer: how does it compare with the CH_4/CH_4 and CF_4/CF_4 dimers? Mol. Phys. **110**, 377–387 (2012)

Chapter 6
Modulus of Elasticity

From the spectroscopic data accumulated in this work, it is possible to extract important properties of the popular materials polyethylene and polytetrafluoroethylene, namely a limiting value for their elastic moduli or Young moduli. In the literature, one finds numerous determinations of this property [1–17], but determinations from isolated molecules, which allow to derive values unaffected by intermolecular interactions, are still rare [18]. Young's modulus (E) is defined for homogeneous materials and small displacements and describes how much stress (force per area, F/A) is necessary to stretch a material by a certain fraction of its length (strain, $\Delta l/l$):

$$E = \frac{\text{stress}}{\text{strain}} = \frac{F/A}{\Delta l/l}. \tag{6.1}$$

The elastic modulus is often given in GPa $= 10^9\,\mathrm{N\,m^{-2}}$, for example $E \approx 200\,\mathrm{GPa}$ for steel, or $E \approx 70\,\mathrm{GPa}$ for aluminum [19].[1] In Fig. 6.1, the definition of Young's modulus is depicted using the example of a rectangular rod. The picture also indicates how this property is connected to the chain molecules investigated in this work: thinking of a carbon chain as a very small "nano-rod", one may ask how much force is necessary to stretch it by a certain amount, or "What is its elastic modulus?". This question is somewhat flawed by the fact that the elastic modulus is defined for macroscopic systems and not directly transferable to molecular systems without ambiguity. However, with a consistent and elaborate approach, one can come to a meaningful and applicable result.

The connection to the spectroscopic data from this work is provided by the fact that longitudinal vibrations of extended carbon chains can be viewed as vibrations of homogeneous rods in the limit of high chain length, when the sequential structure

[1] To set these quantities in relation, one can think of 1 kg pulling on a $1\,\mathrm{cm} \times 1\,\mathrm{cm} \times 10\,\mathrm{cm}$ rectangular rod on the surface of earth with a weight force of $\approx$10 N, which results in an elongation by $\approx$0.01 mm for $E = 1$ GPa. About 200 kg would be needed to elongate a steel-rod by 0.01 mm, or 70 kg in case of an aluminum-rod.

N.O.B. Lüttschwager, *Raman Spectroscopy of Conformational Rearrangements at Low Temperatures*, Springer Theses, DOI 10.1007/978-3-319-08566-1_6

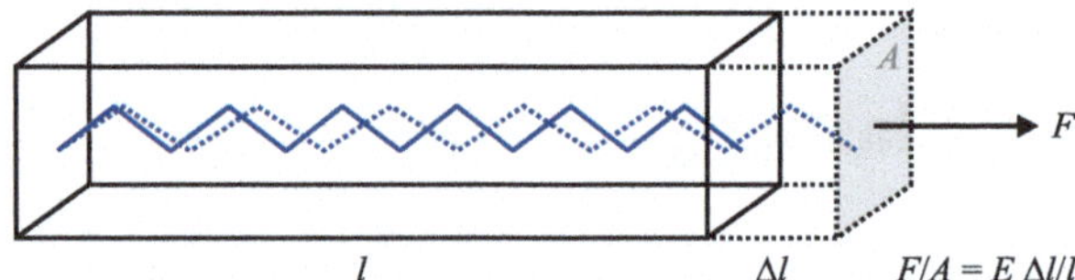

Fig. 6.1 Definition of Young's modulus

Table 6.1 Crystallographic data of all-trans polyethylene and all-anti polytetrafluoroethylene (13/6 helix, form II) from the literature

		Polyethylene [21]	Polytetrafluoroethylene [23]
Segment separation	d/nm	0.1267	0.1300
Cross-section area per molecule	A/nm^2	0.1824	0.2725
Crystal density	$\rho/\mathrm{kg\,dm}^{-3}$	1.008[a]	2.344
Circular cross-section area	A'/nm^2	0.1552	0.2454
Single chain density	$\rho'/\mathrm{kg\,dm}^{-3}$	1.185	2.603

The crystal density is calculated from unit cell parameters, the cross-section area perpendicular to the chains' long axes is calculated from the unit cell's face-area per chain (A) or by assuming a circular cross-section corresponding to cylindrical chains (A'), and the single chain density is calculated from Eq. 6.3

[a]Here, $\rho = 1.00\,\mathrm{kg\,dm}^{-3}$ is used to calculate the elastic modulus, see text. With Eq. 6.3, this translates to $A = 0.1838\,\mathrm{nm}^2$

of the carbon backbone becomes negligible [20]. The physical length of a molecular rod is calculated from the chain length n via:

$$l = nd + c$$

where d is the segment separation of two CH_2 or CF_2 units in direction of the long chain axis and the constant c accounts for terminal hydrogens or fluorines (these can be neglected for long chains). d is available from crystallographic data [21–23] or quantum chemical calculations, which are quite robust in this aspect.[2] Values for this parameter are summarized in Table 6.1.

Longitudinal vibrations of homogeneous rods have wavenumbers [24]:

$$\tilde{\nu} = \frac{m}{2c_0 l}\sqrt{\frac{E}{\rho}}, \tag{6.2}$$

with ρ being the density of the material, c_0 the speed of light in vacuum, and m the number of nodes of the longitudinal mode. Provided that the accordion vibrations of the studied (perfluoro-)alkanes correspond to pure longitudinal vibrations, this relation allows to derive the elastic modulus. To reach the homogeneous rod limit,

[2] The quantum chemical separation is obtained from the slope of a plot of the chain length versus the end-to-end distance, as depicted in Appendix A.7, Figs. A.12 and A.13.

the acquired data must be extrapolated to infinite chain length, so that a value for the polymeric counterparts of the oligomers investigated here will be obtained.

The density occurring in Eq. 6.2 is a substantial source of ambiguity when it comes to derive the elastic modulus from investigations on the molecular level. The true source of ambiguity is actually not the density, but the cross-section area (A) occurring in Eq. 6.1, that mutually depends on the density. For carbon chains, this mutual dependence may be expressed as:

$$\rho = \frac{M(\mathrm{CX_2})}{N_A\, d\, A}, \tag{6.3}$$

where N_A denotes Avogadro's constant, and $M(\mathrm{CX_2})$ is the molar mass of a chain segment (X = H, F). Problems arise because there are no sharp molecular boundaries, and therefore no strictly defined cross-section area, which is why an *effective* cross-section area needs to be employed in the calculation of the elastic modulus. It is the arbitrariness in the definition of such an effective cross-section area which has led to some confusion and limited comparability of elastic moduli reported in the literature, especially in case of polytetrafluoroethylene.

For polyethylene, the most recent experimental determinations of the elastic modulus [11, 18] employed the density $\rho = 1.00\,\mathrm{kg\,dm^{-3}}$, which will be used in the following calculations to establish comparability. This value is the approximate crystal density obtained from the evaluation of an X-ray structure determination published by Bunn [21]. From Bunn's crystal structure follows an effective cross-section area $A = 0.1824\,\mathrm{nm^2}$ when the basal face-area of the unit cell is uniformly apportioned to the number of chains per unit cell (denoted 'unit cell face-area per chain' from here on). With Eq. 6.3, this area actually translates to $\rho = 1.008\,\mathrm{kg\,dm^{-3}}$ ($\approx 1.01\,\mathrm{kg\,dm^{-3}}$), the value listed in Table 6.1. The unit cell face-area per chain $A = 0.1824\,\mathrm{nm^2}$ was used in several computational investigations [2, 3, 5] to convert the calculated quantities into the elastic modulus. Therefore, experimental and computational studies work with slightly different underlying values causing a small deviation of the order of 1 %.

Studies on the elastic modulus of polytetrafluoroethylene defined the effective cross-section area differently. Computational studies report varying A-values ($A = 0.253\,\mathrm{nm^2}$ [12], $A = 0.2484\,\mathrm{nm^2}$ [13, 14]) incompatible with the unit cell face-area per chain calculated from the crystal structure [23] ($A = 0.2725\,\mathrm{nm^2}$). Apparently, these studies implied a circular cross-section based on the assumption of cylindrical chains, excluding any blank volume in the crystal from the effective cross-section area. Approximating the chains as cylinders lying on the long edges of the pseudo-hexagonal unit cell (edge-length $a = 0.559\,\mathrm{nm}$ [23]) gives a circular cross-section area $A' = \pi(a/2)^2 = 0.245\,\mathrm{nm^2}$, closer to the values listed above. This implies that these studies did not work with the crystal density, $\rho = 2.344\,\mathrm{kg\,dm^{-3}}$ [23], but rather with significantly larger "single chain densities" $\rho' = 2.52\,\mathrm{kg\,dm^{-3}}$ and $\rho' = 2.57\,\mathrm{kg\,dm^{-3}}$.[3]

[3] Calculated using Eq. 6.3 with $M(\mathrm{CF_2}) = 50.0\,\mathrm{g\,mol^{-1}}$.

This introduces two ways of defining the effective cross-section area of poly(tetrafluoro)ethylene chains. A third one is to approximate the outer boundary of the chains by van-der-Waals radii [14] (Fig. A.14, Appendix A.7). However, van-der-Waals radii are inherently somewhat arbitrary, and form a poor basis for this definition. In principle, the other two definitions are both applicable, but have a different focus. Using the crystal density (and thus an effective cross-section area given by the unit cell face-area per chain) yields an ultimate elastic modulus for close-packed chains, while a circular cross-section area (and thus effectively a single chain density) yields the elastic modulus of a single "nano-rod". The latter will be larger than the ultimate crystal modulus, because the crystal unavoidably contains some blank volume. From the perspective of the jet-isolation experiment, it would be consistent to give the nano-rod modulus, but comparability with literature values demands to employ the crystal density as well. For polytetrafluoroethylene, literature moduli will be adjusted to the density and cross-section area used here to establish comparability. Densities and corresponding cross-section areas are listed in Table 6.1.

6.1 LAM-1 Perturbation

With a reasonable choice for length and density, the wavenumbers of longitudinal chain stretching vibrations need to be found. The accordion vibrations assigned in Chaps. 4 and 5 correspond to pure longitudinal chain stretching vibrations with a few exceptions. Earlier in the text it was stated that the accordion vibration mixes with transversal acoustic vibrations at certain chain lengths, yielding vibrations with partial transversal and longitudinal displacement character. If this is the case, the concerned vibration must be "purified" from non-longitudinal contributions, before using it to extrapolate the elastic modulus. The reason for the mode-mixing is to be sought in the form of the frequency branch to which the accordion vibrations belong. The detailed discussion of this issue will be restricted to polyethylene [20], but the corresponding frequency branch of polytetrafluoroethylene [25] has a similar shape, so that the argumentation is applicable in this case as well.

For polyethylene, the accordion vibration belongs to the C–C–C bending frequency branch, which is depicted in Fig. 6.2 (derived from B3LYP calculations, see Appendix A.8). Modes at low phase values are characterized by longitudinal displacements (LAMs), modes at higher phase values by transversal displacements (TAMs) [20]. Because the C–C–C bending frequency branch is an acoustical one with zero-frequency at limiting phase values, it necessarily passes through a maximum. Modes of finite alkanes are uniformly distributed over this frequency branch, indicated by blue and red dots in Fig. 6.2 (the color denotes the particular symmetry). At certain chain lengths, some modes on opposite sites of the maximum will have similar frequencies. If the frequency and symmetry matches, normal modes of the finite system are mixtures of the deperturbed LAM and TAM. In each example depicted in the figure, the frequency of the LAM-1 falls in close proximity to the

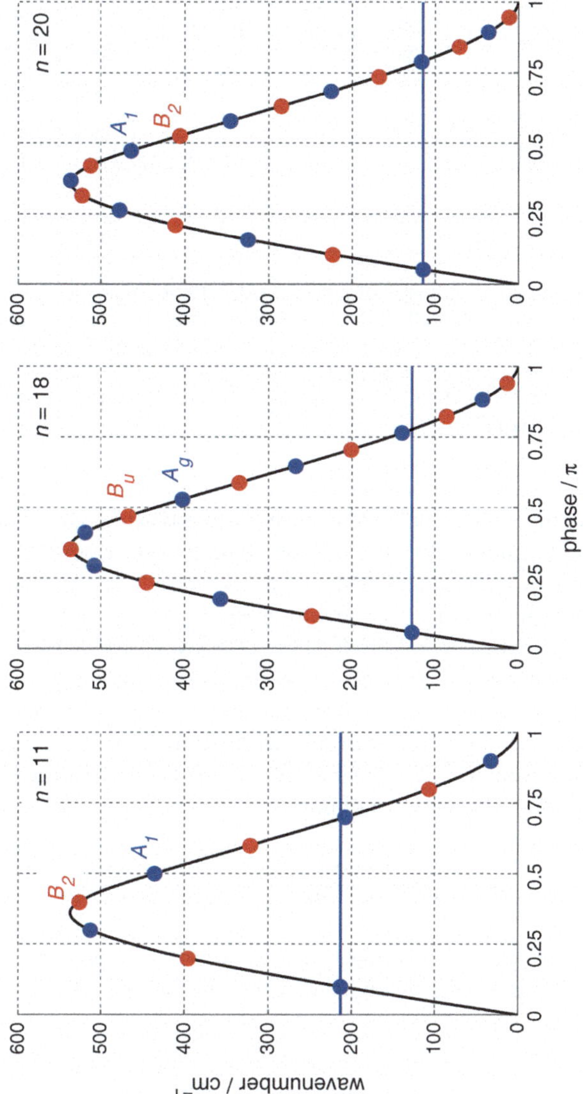

Fig. 6.2 Distribution of modes over the polyethylene C–C–C bending frequency branch for $n =$ 11, 18, and 20. *Colors* indicate the symmetry of the modes: *blue* $= A_{1(g)}$, *red* $= B_{2(u)}$. The LAM-1 is the totally symmetric (*blue*) mode with the lowest phase value. Perturbation of the LAM-1 occurs when its frequency (highlighted by a *blue line*) comes close to the frequency of a totally symmetric mode at the high-phase value end of the branch

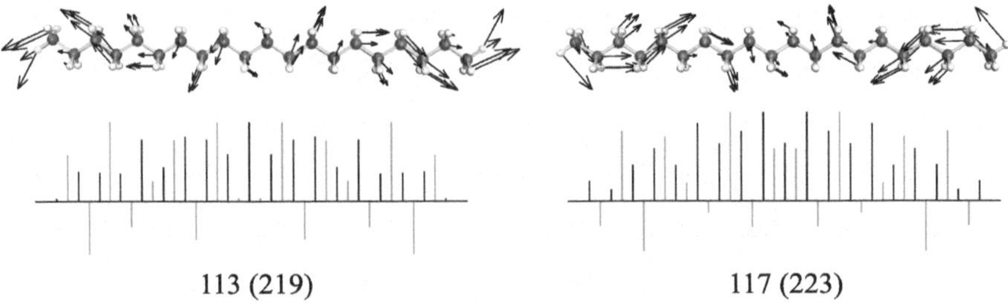

Fig. 6.3 Harmonic normal coordinates of mixed LAM-1/TAM vibrations of eicosane calculated with Turbomole v6.4 [26] on the B3LYP-D3/6-311++G** level. Unscaled wavenumbers in cm^{-1} and the scattering cross-section (in parentheses, $T_{\text{vib}} = 100\,\text{K}$) in $10^{-36}\,\text{m}^2\text{sr}^{-1}$ are given at the *bottom*. Relative internal coordinate displacements are drawn below the molecular structures (C–C stretching in *black*, C–C–C bending in *gray*, see Fig. 4.6). The deperturbed LAM-1 wavenumber, calculated with Eq. 6.4, is $115\,\text{cm}^{-1}$

frequency of a totally symmetric TAM at the high-phase value end of the branch, leading to mixed normal modes. In case of other chain lengths from 10 to 21 (Fig. A.16 in Appendix A.8), mixing of the LAM-1 is smaller or negligible. The closest match of wavenumbers is found for eicosane with $n = 20$, where the LAM-1 exhibits the strongest coupling on the B3LYP-D3/6-311++G** level for $n = 13$–21. The strongly mixed normal coordinates of this special case are shown in Fig. 6.3.

The "purification" of accordion vibrations and recovery of deperturbed LAM-1 frequencies can be accomplished by using perturbation theory. Treating the perturbations like a Fermi resonance [27, 28] and calculating the deperturbed LAM position from the experimental intensity weighting:

$$\tilde{\nu}^0_{\text{LAM-1}} = \frac{\tilde{\nu}_{\text{LAM-1}} I_{\text{LAM-1}} + \tilde{\nu}_{\text{TAM}} I_{\text{TAM}}}{I_{\text{LAM-1}} + I_{\text{TAM}}} \tag{6.4}$$

was shown to work well to this end [18, 29, 30]. This approach implies that the perturber has negligible intensity on its own, which is well realized in case of alkane and perfluoroalkane TAMs.

6.2 Extrapolation to Infinite Chain Length

If the investigated all-trans alkanes and all-anti perfluoroalkanes behave like homogeneous rods, their LAM-1 wavenumbers should satisfy Eq. 6.2 and be antiproportional to the chain length. In terms of the corresponding frequency branches [25, 31], this means that their LAMs need to fall in the range of low phase values, where the frequency dispersion is essentially linear. Since the phase of LAM-1 vibrations is inversely proportional to the chain length, this is true in case of sufficiently long molecules. Also, only in this limit, deviations due to end groups will become insignificant.

Following from Eq. 6.2 with $l = nd$, the LAM-1 wavenumber (abbreviated $\tilde{\nu}$ in the following) can then be expressed as:

$$\tilde{\nu} = 1/n \cdot \underbrace{\frac{1}{2c_0 d}\sqrt{\frac{E}{\rho}}}_{=S}, \tag{6.5}$$

so that the elastic modulus may be obtained from the slope S of a plot of LAM-1 wavenumbers against the inverse chain length:

$$E = 4\rho(c_0 d S)^2. \tag{6.6}$$

The quantity which corresponds to infinite chain length is the initial slope of the frequency branch. If one plots and linearly fits LAM-1 wavenumber of insufficiently long chains, the increasing negative curvature of the frequency branch with increasing phase value will lead to a too small estimate of the initial slope. The derived elastic modulus will correspond to a lower bound. For alkanes, which were measured up to $n = 21$, it will be seen that the measured chains were long enough to allow an extrapolation based on Eq. 6.5. In case of perfluoroalkanes however, the data is limited to $n \leq 14$, and deviations from Eq. 6.5 are obvious.

In order to obtain the initial slope of the polytetrafluoroethylene frequency branch from the rather restricted data set, it is thus necessary to extend Eq. 6.5. The extrapolation approach

$$n\tilde{\nu} = S + \frac{B}{(n + C)^2} \tag{6.7}$$

was suggested in Ref. [18], and will be applied here as well. This slightly rearranged version of Eq. 6.5 uses an additional term defined by two parameters B and C to account for higher order deviations from linearity in $1/n$ at short chain length. The parameter C is not varied freely, but optimized to linearize a plot of $n\tilde{\nu}$ against $(n + C)^{-2}$ from a large set of LAM-1 wavenumbers, calculated using low-level but far-reaching quantum chemistry. With a properly chosen parameter C, the effect of restricting the data to short chain lengths is largely attenuated. S, the y-intercept of the $n\tilde{\nu}$ versus $(n + C)^{-2}$ plot, is extrapolated from experimental LAM-1 wavenumbers using the quantum chemically calibrated parameter C. Subsequent utilization of Eq. 6.6 yields a semi-empirical elastic modulus which will be reviewed carefully. Fortunately, the more comprehensive data set for alkanes allows the evaluation of this extrapolation technique.

Table 6.2 Averaged wavenumbers of normal alkane LAM-1, perturbing TAM, and LAM-3 bands from Raman jet measurements in cm^{-1}

Chain length	Original (TAM, FR) LAM-1	$I_{TAM, FR}/I_{LAM-1}$	Deperturbed LAM-1	LAM-3
13	(149.0) 176.0	0.1382	172.8	444
14	(158.0) 165.2	0.6388	162.4	433
15	154.4	—	—	398
16	143.3	—	—	389
17	135.6	—	—	377
18	126.8	—	—	352
19	121.0	—	—	341
20	(109.9) 115.7	0.1370	115.0	321
21	109.1	—	—	310

LAM-1 and TAM positions and intensities were determined by fitting Gaussian curves, LAM-3 positions are peak wavenumbers. In case of tridecane, tetradecane (Fermi resonance), and eicosane, the deperturbed LAM-1 wavenumber was calculated using Eq. 6.4

6.2.1 Extrapolation of Normal Alkane LAM-1 Wavenumbers

Deperturbed wavenumbers of normal alkane LAM-1 bands, assigned in Sect. 4.4, are summarized in Table 6.2, together with untreated wavenumbers of LAM-3 bands. Deperturbation was considered in case of tridecane, tetradecane, and eicosane. The tetradecane LAM-1 vibration is perturbed not by a fundamental TAM level, but by an anharmonic resonance with a combination tone of two TAMs [18]. A TAM perturbation would be incompatible with the C–C–C bending frequency branch, as shown in Fig. A.16 (Appendix A.8). For octadecane, the TAM predicted to perturb the LAM-1 is obscured by single gauche bands and cannot be measured. In the remaining cases, the intensities and band positions of LAM-1 bands and perturbers are determined by fitting Gaussian curves. The position of the deperturbed LAM-1 is recovered using Eq. 6.4. LAM-3 bands were not fitted, since they deviate too much from the homogeneous rod approximation (see below). The accuracy of wavenumbers is improved by double-checking the calibration of the spectra using rotational N_2 transitions [32] and CF_4 vibrational bands [33, 34]. Accounting for uncertainties in measuring band centers and weighting in case of perturbed LAMs, the wavenumber uncertainty is estimated to be not more than $\pm 1\,cm^{-1}$.

In Fig. 6.4, the wavenumbers of LAM-1 modes are plotted against the inverse chain length and fitted in the range $n = 16$–21 using Eq. 6.5. At these chain lengths, the data does not exhibit a systematic deviation from a linear curve anymore, as the residual shows. The pentadecane LAM-1 lies above the fitted curve indicating a missed perturbation (judging from the C–C–C frequency branch, Fig. A.16, rather a Fermi resonance than a mixing TAM), and is omitted. Tetradecane and tridecane seem to be too short to let their LAM-1 fall on the initial linear portion of the frequency branch. The fitted slope is $S = (2295 \pm 8)\,cm^{-1}$. The systematic uncertainty from

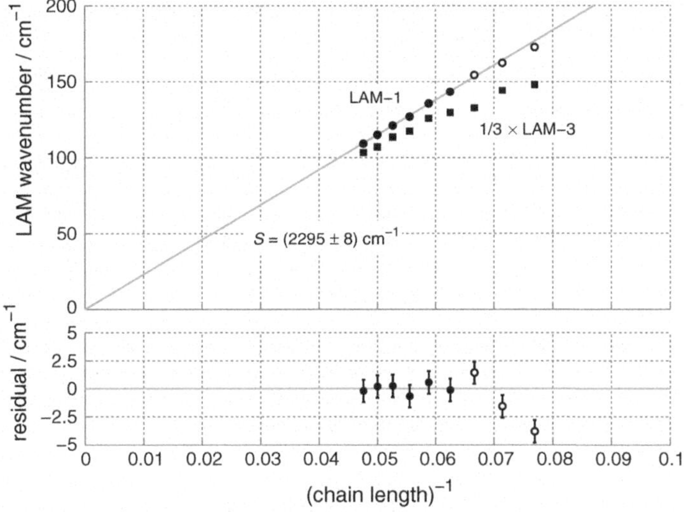

Fig. 6.4 Extrapolation of Raman jet LAM-1 wavenumbers (*circles*) of normal alkanes (n = 13–21) to infinite chain length using Eq. 6.5. Tridecane, tetradecane and pentadecane are excluded from the fit (*open circles*). LAM-3 wavenumbers (*squares*) should fall on LAM-1 wavenumbers when divided by their number of nodes, if the elastic homogeneous rod approximation holds. This is not yet the case in the size range n = 13–21

restricting the data to $n \leq 21$ is probably not more than the uncertainty from scattering. Conservatively adding the uncertainties yields $S = (2.30 \pm 0.02) \times 10^3\,\mathrm{cm}^{-1}$.

Wavenumbers of LAM-3 modes are also included in Fig. 6.4. The applied division by the number of nodes, $m = 3$, should let them coincide with LAM-1 modes in the limit of high n (see Eq. 6.2). As in case of LAM-1 modes, this requires them to fall in the range of phase values where the C–C–C bending frequency branch is essentially linear, which is apparently not the case for n = 13–21. Therefore, only LAM-1 modes allow to extrapolate the elastic modulus using Eq. 6.2.

6.2.2 Extrapolation of Perfluoroalkane LAM-1 Wavenumbers

The assignment of perfluoroalkane LAM-1 vibrations and perturbing vibrations was discussed in Chap. 5 and is listed in Table 6.3. The band positions and intensities were obtained by fitting Gaussian curves and transfered to the deperturbed LAM-1 wavenumber using Eq. 6.4. Other than in case of normal alkanes, the spectra lack rotational lines from air (except the $n = 14$ spectrum, which was measured with the less air-tight heatable brass saturator) or bands from other compounds which could be used to validate the wavenumber calibration. Therefore, small calibration corrections up to 1.2 cm^{-1} were made based on the assumption that the wavenumber of the CF_2

Table 6.3 Wavenumbers of perfluoroalkane LAM-1 bands and bands from perturbing TAMs (in parentheses) from He Raman jet measurements in cm^{-1}

Chain length	Original (TAM) LAM-1	$I_{TAM}/I_{LAM\text{-}1}$	Deperturbed	Calibration-corrected
6	(141.9) 168.3	0.7479	157.0	157.1
8	(116.8) 131.4	0.2969	128.0	127.7
9	115.7	—	—	114.8
10	(101.5) 108.3	0.2568	106.9	106.1
12	(84.0) 90.2	0.1251	89.5	90.5
13	82.9	—	—	84.1
14	79.8	—	—	79.4

Band positions and intensities were determined by fitting Gaussian curves. Perturbation was removed in case of $n = 6, 8, 10$, and 12 using Eq. 6.4. In order to minimize scattering due to shortcomings in the wavenumber calibration, the calibration was corrected assuming the position of the CF_2 scissoring vibration is chain length independent, see text

scissoring vibration ($\approx$385 cm^{-1}) is chain length independent. The experimental and computational wavenumbers of this vibration show no systematic evolution with chain length, but rather scatter slightly around the average value. The experimental average was used as reference point.

A plot of the deperturbed and calibration-corrected LAM-1 wavenumbers against the inverse of the chain length is shown in Fig. 6.5a. It is obvious that the measured perfluoroalkanes are too short to satisfy Eq. 6.5 and thus do not fall on a straight line. Using Eq. 6.5, one can only calculate a lower bound for the initial slope of the frequency branch from the LAM-1 of perfluorotetradecane, giving $S > 1.11 \times 10^3$ cm^{-1} (dash-dotted line). Accounting for higher order deviations but avoiding reference to quantum chemical predictions for the moment is feasible by fitting higher order polynomials [9]. Fitting a second order polynomial yields an initial slope of $S = (1.24 \pm 0.02) \times 10^3$ cm^{-1}, but it is unlikely that such a limited function already describes the deviations from linearity sufficiently accurate. The dashed line in Fig. 6.5a shows a fit of a third order polynomial, which describes the data set better, giving $S = (1.22 \pm 0.09) \times 10^3$ cm^{-1}. However, fitting three parameters to seven pairs of values leads to higher uncertainty. Using the extrapolation based on Eq. 6.7 and substituting one otherwise free parameter by a quantum chemical estimate goes beyond second order deviations without raising the uncertainty so much.

As introduced above, the quantum chemical estimate is given by the parameter C. Optimization to perfluorinated alkanes was carried out using harmonic frequency calculations on the HF/STO-3G level [35] performed with Gaussian 03 [36]. Equation 6.7 was found to linearize a data set in the size range $n = 6$–50 best when C was set to 6.5 (Fig. A.23, Appendix A.9). This value was tested by application to smaller B3LYP/6-311+G(d) [35] and B3LYP-D3/def2-TZVP data sets, yielding well linear curves which can straightforwardly be extrapolated to infinite chain length (Figs. A.23 and A.24, Appendix A.9). The quantum chemical investigations also revealed a LAM-1 wavenumber alternation depending on the chain length being an even or odd number, which is magnified by the multiplication of the

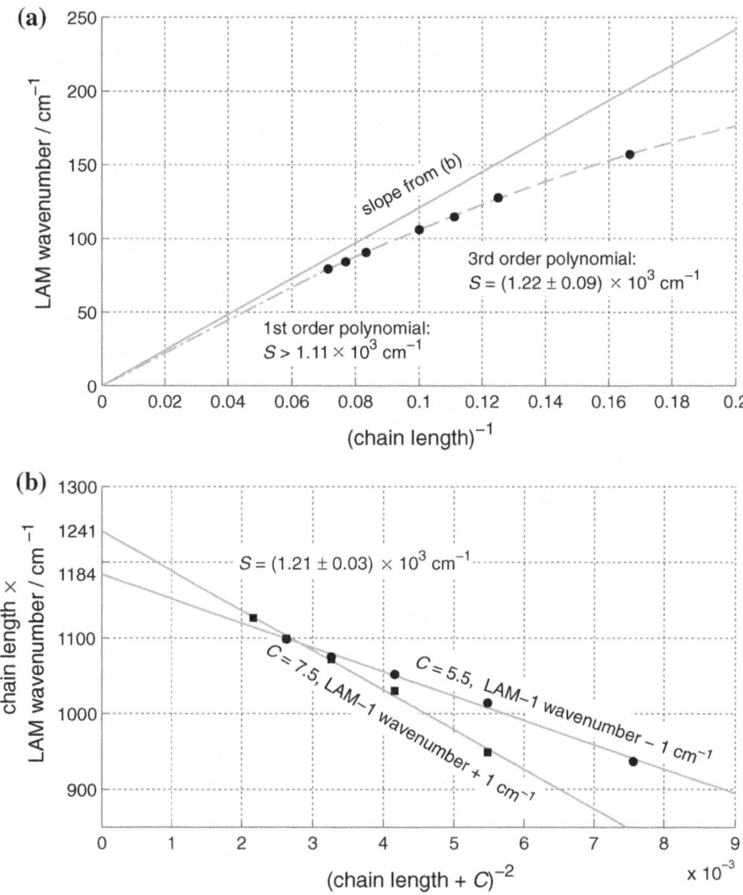

Fig. 6.5 Extrapolation of Raman jet LAM-1 wavenumbers of perfluoroalkanes to infinite chain length using Eq. 6.5 (**a**) and Eq. 6.7 (**b**). Reproduced and adapted material from reference III

LAM-1 wavenumber with the corresponding chain length. Even- and odd-numbered chain lengths were thus evaluated separately, but without variation of C. According to the far-reaching HF calculations, this ramification of sequential structure ceases only at high chain length about $n \approx 50$. The fits for odd and even chain length converge in case of the HF calculations, as one would expect for chains long enough to behave like a continuous medium, and the alternation does thus not affect the obtained y-intercept. However, this is not true in case of the DFT data sets, where the linear fits run rather parallel. By varying C, DFT-curves for even and odd chain lengths can be adjusted to converge to a single y-intercept, but only at unreasonable expense of linearity and/or at unreasonably high C ($C = 55$ in case of the B3LYP-D3/def2-TZVP, for example). The small DFT data sets can thus not be used to evaluate C. Rather, one has to rely on the HF prediction and accept some uncertainty.

The $n\tilde{\nu}$ versus $(n+C)^{-2}$ plot of experimental data is shown in Fig. 6.5b. The two available LAM-1 wavenumbers of perfluoroalkanes with odd chain length are not enough for a reliable extrapolation and are thus omitted. An upper bound for S is established by fitting Eq. 6.7 setting $C = 7.5$ and raising the LAM-1 wavenumber by $1\ \mathrm{cm}^{-1}$, a lower bound by setting $C = 5.5$ and lowering the LAM-1 wavenumber by $1\ \mathrm{cm}^{-1}$, accounting for an uncertainty of ± 1 ($\approx$15 %) in C and $\pm 1\ \mathrm{cm}^{-1}$ ($\approx$1 %) in $\tilde{\nu}$. This leads to $S = (1.21 \pm 0.03) \times 10^3\ \mathrm{cm}^{-1}$, which almost reproduces the result of fitting a third order polynomial to the LAM-1 wavenumber plotted against the inverse chain length, but reduces the uncertainty by a factor of three. The uncertainty of this semi-empirical result increases from 2 to 3 % if one allows a variation of C by ± 2 ($\approx$30 %), which is still less than half the uncertainty involved in the third order polynomial fit (7 %). A much larger uncertainty in C would be at variance with the fact that the semi-empirical approach worked quite well in case of shorter alkanes [18] (see below), where rather crude HF/3-21G calculations were employed as well. The involvement of quantum chemistry is thus indeed helpful, despite the restriction to low computational levels.

6.3 Comparison of Elastic Moduli

Elastic moduli of polyethylene and polytetrafluoroethylene are calculated from the initial slopes of the corresponding frequency branches (S) with Eq. 6.6, yielding the values listed in Table 6.4. Moduli applicable to crystalline samples and nano-rod values are provided. These values differ solely in the underlying chain cross-section area (Table 6.1), as discussed above.

In order to compare the stiffness of polyethylene and polytetrafluoroethylene chains, it is necessary to come back to one of the initial questions of this chapter: what is the force needed to stretch these chains by a certain amount? The force generating a strain of 1 % is listed in Table 6.4, calculated by plugging the elastic modulus and associated cross-section area into Eq. 6.1. One can see that the calculated force is virtually identical in both cases, in agreement with a previous claim that this quantity hardly depends on groups attached to the carbon chains, and that the small deviation of the torsional angles in all-trans alkane chains and all-anti perfluoroalkane chains

Table 6.4 Elastic moduli of polyethylene and polytetrafluoroethylene calculated from the initial slope of the corresponding frequency branches (S)

	Slope $\times 10^{-3}/\mathrm{cm}^{-1}$	Crystal/GPa	Nano-rod/GPa	Force/pN
Polyethylene	2.30 ± 0.02	305 ± 5	362 ± 6	$(5.6 \pm 0.1) \times 10^2$
Polytetrafluoroethylene	1.21 ± 0.03	209 ± 10	232 ± 12	$(5.7 \pm 0.3) \times 10^2$

The calculations are made using crystal densities and theoretical single chain densities, assuming a circular chain cross-section area (column ‘nano-rod’). The last column is the force necessary to stretch an extended isolated PE or PTFE chain by 1 %, calculated with Eq. 6.1. ρ- and A-values are listed in Table 6.1

has no significant influence on the corresponding stiffness [15]. The different elastic moduli of polyethylene and polytetrafluoroethylene are merely an effect of different chain-packing, that is, more chains per unit cross-section in polyethylene.

Another useful comparison of the forces given in Table 6.4 is to set them against the force needed to stretch a single C–C bond. This force may be estimated from the C–C stretching vibrational wavenumber of ethane ($\tilde{\nu} = 995\,\mathrm{cm}^{-1}$ [37]). Calculation with a reduced mass of $\mu \approx 7.5\,\mathrm{g\,mol}^{-1}$ (assuming the methyl groups move as rigid masses) yields the force constant $k \approx 440\,\mathrm{N\,m}^{-1}$ which translates to a force of $f \approx 670\,\mathrm{pN}$.[4] This force is larger than in case of a (coiled) zig-zag chain, because deformation of the latter involves softer C–C–C bending (and C–C–C–C torsion) coordinates, along with stiffer C–C stretching coordinates.

The determined elastic moduli calculated from crystal densities are set against literature values in Tables 6.5 (polyethylene) and 6.6 (polytetrafluoroethylene). From the perspective of the preparation technique, the extrapolated moduli are idealized values for isolated chains in vacuum at low-temperatures. This needs to be considered when comparing the Raman jet-results to literature moduli. Therefore, the most comparative value in case of polyethylene is the result from the previous Raman jet extrapolation for alkanes with $n = 5$–16, which yielded $E = (309 \pm 8)\,\mathrm{GPa}$ [18]. Within the accuracy of the measurements, both values are virtually identical. This underlines the validity of the extrapolation approach employed in Ref. [18] and strengthens the confidence in the accuracy of the semi-empirical value obtained here for polytetrafluoroethylene.

Further comparison of polyethylene moduli was discussed in Refs. [18] and [30] and shall be summarized just briefly. The deviation of the obtained value to elastic moduli of amorphous polyethylene samples is in particular noticeable. Common samples have elastic moduli orders of magnitudes smaller than the vacuum limit, of the order of 0.1 GPa [38]. High crystallinity is necessary to obtain samples with moduli closer to the Raman jet value [39, 40], which can reach values up to $\approx 290\,\mathrm{GPa}$ [8]. Raman spectroscopic investigations of solid alkanes yielded a limiting modulus of $(305 \pm 3)\,\mathrm{GPa}$, after corrections for intermolecular interactions were considered, which agrees perfectly with Raman jet values. Without such corrections, a significantly higher value of $(364 \pm 25)\,\mathrm{GPa}$ is extrapolated from solid alkane LAM-1 wavenumbers [9].

More direct measurements follow microscopic strain in semi-crystalline polyethylene by measuring variations in lattice plane distances using X-ray diffraction [7, 8]. The modulus is calculated from the macroscopic stress applied to the measured sample, assuming the stress is uniformly distributed over the sample, which is a controversial point of discussion [41] and a potential explanation for the X-ray values usually being smaller than those from spectroscopic determinations. The same problem concerns stress-strain measurements of polytetrafluoroethylene samples [15], although a recently reported modulus derived from neutron diffraction measurements [16] even exceeds the value obtained here.

---

[4] $k = \mu\,(2\pi\,c\,\tilde{\nu})^2$; $f \approx k \times 0.153\,\mathrm{pm} \times 1\,\%$.

Quantum chemical studies often convert a calculated change of energy per strain to the elastic modulus. Values from HF [1, 2] and DFT calculations [5, 6] tend to be somewhat too large, while a reported MP2 calculation almost matches the modulus reported here [3]. The Raman-jet value from this work and the previous jet-investigation invite for further quantum chemical studies, putting thermal effects [4] and interchain interactions [6] aside.

Comparing elastic moduli of polytetrafluoroethylene needs more attention to ensure transferability. One issue are varying densities or cross-section areas, as discussed in the beginning of this chapter. Literature values listed in Table 6.6 were unified using densities reported by Clark [23]. Another issue which needs to be considered is the structure of the investigated samples. Most of the literature moduli are determined from 15/7 helical polytetrafluoroethylene (form IV), at variance with the 13/6 helical conformation of jet-isolated perfluoroalkanes. However, the above calculated forces needed to stretch polyethylene and perfluoropolyethylene chains suggests that this small structural deviation should not influence the elastic modulus significantly. Deviations should mainly stem from the slightly different densities of 13/6 and 15/7 helical polytetrafluoroethylene [23].

Unfortunately, an important reference reporting experimental moduli [17] (including data from Ref. [25]) lacks the information which density was used in the underlying calculations. However, the respective publication includes a plot of LAM-1 wavenumbers against the inverse chain length, from which the slope $S \approx 1150\,\mathrm{cm}^{-1}$ can be estimated. The reported E-value (206 GPa) then implies a single chain density

Table 6.5 Elastic modulus of polyethylene (in GPa), determined by different methods

Reference	Method/physical condition	Reported modulus
Computational		
[1]	HF calculation	380 ± 50
[2]	HF calculation	345
[3]	MP2 calculation	303
[4]	Molecular mechanics simulation	318
[5]	DFT calculation	334
[6]	DFT calculation	360
X-ray diffraction, stress-strain measurement		
[7]	Solid state	235
[8]	Solid state	288 ± 10
Raman spectroscopy, LAM frequency extrapolation		
[9]	Solid state	364 ± 25^a
[10]	Solid state^b	290
[11]	Solid state^b	305 ± 3
[18]	Jet-isolated	309 ± 8
This work	Jet-isolated	(305 ± 5)

^{a}converted using $\rho = 1.00\,\mathrm{kg\,dm}^{-3}$
^{b}corrected for intermolecular interactions

Table 6.6 Elastic modulus of polytetrafluoroethylene (in GPa), determined by different methods

Reference	Structure	Method (physical condition)	Reported modulus		Unified[a]
Computational					
[12]	13/6	Force field calculation	160	$A = 0.253\,nm^2$	150
[13]	13/6	DFT calculation	221	$A = 0.2484\,nm^2$	201
[14]	planar	DFT calculation	247	$A = 0.2484\,nm^2$	226
Experimental					
[15]	15/7	X-ray diffraction, stress-strain measurement (solid)	153		
[16]	15/7	Neutron diffraction, stress-strain measurement (solid)	≈220		
[17, 25]	15/7	Neutron scattering, mapping of phonon dispersion curve (solid)	222	$\rho \approx 2.56\,kg\,dm^{-3}$	≈200
[17]	15/7	Raman spectroscopy, LAM frequency extrapolation (solid, liquid)	206	$\rho \approx 2.56\,kg\,dm^{-3}$	≈ 185, (201)[b]
This work	13/6	Raman spectroscopy, LAM frequency extrapolation (jet-isolated)			(209 ± 10)

For comparability, literature values are translated with densities (ρ) or cross-section areas (A) used in this work (column 'unified'), if applicable. Original ρ- and A-values are given in these cases

[a] 13/6 helix: $\rho = 2.344\,kg\,dm^{-3}$, $A = 0.2725\,nm^2$; 15/7 helix: $\rho = 2.302\,kg\,dm^{-3}$, $A = 0.2774\,nm^2$ [23]

[b] revised here, see text

of $\approx 2.56\,\mathrm{kg\,dm^{-3}}$. With the recent crystal density of 15/7 helical polytetrafluoroethylene, $\approx 2.302\,\mathrm{kg\,dm^{-3}}$ [23], the slope rather yields $E = 185\,\mathrm{GPa}$. This modulus is somewhat lower than the modulus obtained here, but closer examination of the published plot reveals that the slope was actually underestimated. Using just the LAM-1 wavenumber of the longest perfluoroalkane reported in Ref. [17], $\tilde{\nu} = 60\,\mathrm{cm^{-1}}$ ($n = 20$), the modulus $E = 201\,\mathrm{GPa}$ is calculated using Eq. 6.2 ($S = 1200\,\mathrm{cm^{-1}}$), in much better agreement with the Raman jet value. Other than in case of polyethylene [18], intermolecular interactions between lamella of extended chains in the crystal do not seem to affect the elastic modulus of polytetrafluoroethylene significantly, so that no correction is necessary [10, 11]. Calculating with a unified density (or cross-section area) also brings the value obtained from neutron scattering [25] and moduli obtained from DFT calculations [13, 14] in closer agreement with the result from this work. Then, it seems there is consensus that the elastic modulus of crystalline polytetrafluoroethylene lies in the range 200–220 GPa.

References

1. A.L. Brower, J.R. Sabin, B. Crist, M.A. Ratner, Ab initio molecular orbital studies of polyethylene deformation. Int. J. Quantum Chem. **18**, 651–654 (1980)
2. A. Karpfen, Ab initio studies on polymers. V. All-trans-polyethylene. J. Chem. Phys. **75**, 238–245 (1981)
3. S. Suhai, Ab initio calculation of polyethylene deformation including electron correlation effects. J. Polym. Sci. Polym. Phys. Ed. **21**, 1341–1346 (1983)
4. D.J. Lacks, G.C. Rutledge, Simulation of the temperature dependence of mechanical properties of polyethylene. J. Phys. Chem. **98**, 1222–1231 (1994)
5. J.C.L. Hageman, R.J. Meier, M. Heinemann, R.A. de Groot, Young modulus of crystalline polyethylene from ab initio molecular dynamics. Macromolecules **30**, 5953–5957 (1997)
6. G.D. Barrera, S.F. Parker, A.J. Ramirez-Cuesta, P.C.H. Mitchell, The vibrational spectrum and ultimate modulus of polyethylene. Macromolecules **39**, 2683–2690 (2006)
7. I. Sakurada, T. Ito, K. Nakamae, Elastic moduli of the crystal lattices of polymers. J. Polym. Sci. Part C Polym. Symp. **15**, 75–91 (1967)
8. P.J. Barham, A. Keller, The achievement of high-modulus polyethylene fibers and the modulus of polyethylene crystals. J. Polym. Sci. Polym. Lett. Ed. **17**, 591–593 (1979)
9. R.F. Schaufele, T. Shimanouchi, Longitudinal acoustical vibrations of finite polymethylene chains. J. Chem. Phys. **47**, 3605–3610 (1967)
10. G.R. Strobl, R. Eckel, A Raman spectroscopic determination of the interlamellar forces in crystalline n-alkanes and of the limiting elastic modulus E_c of polyethylene. J. Polym. Sci. Polym. Phys. Ed. **14**, 913–920 (1976)
11. R.G. Snyder, H.L. Strauss, R. Alamo, L. Mandelkern, Chain-length dependence of interlayer interaction in crystalline n-alkanes from Raman longitudinal acoustic mode measurements. J. Chem. Phys. **100**, 5422–5431 (1994)
12. T. Shimanouchi, M. Asahina, S. Enomoto, Elastic moduli of oriented polymers. I. The simple helix, polyethylene, polytetrafluoroethylene, and a general formula. J. Polym. Sci. **59**, 93–100 (1962)
13. F. Bartha, F. Bogár, A. Peeters, C. Van Alsenoy, V. Van Doren, Density-functional calculations of the elastic properties of some polymer chains. Phys. Rev. B Condens. Matter Mater. Phys. **62**, 10142–10150 (2000)

14. M.-L. Zhang, M.S. Miao, A. Peeters, C.V. Alsenoy, J.J. Ladik, V.E. Van Doren, LDA calculations of the Young's moduli of polyethylene and six polyfluoroethylenes. Solid State Commun. **116**, 339–343 (2000)
15. I. Sakurada, K. Kaji, Relation between the polymer conformation and the elastic modulus of the crystalline region of polymer. J. Polym. Sci. Part C Polym. Symp. **31**, 57–76 (1970)
16. E.N. Brown, P.J. Rae, D.M. Dattelbaum, B. Clausen, D.W. Brown, In-situ measurement of crystalline lattice strains in polytetrafluoroethylene. Exp. Mech. **48**, 119–131 (2008)
17. J.F. Rabolt, B. Fanconi, Longitudinal acoustic modes of polytetrafluoroethylene copolymers and oligomers. Polymer **18**, 1258–1264 (1977)
18. T.N. Wassermann, J. Thelemann, P. Zielke, M.A. Suhm, The stiffness of a fully stretched polyethylene chain: a Raman jet spectroscopy extrapolation. J. Chem. Phys. **131**, 161108 (2009)
19. *CRC Handbook of Chemistry and Physics*, ed. by D.R. Lide, 82nd edn, (CRC Press, Boca Raton, 2001)
20. T. Shimanouchi, Local and overall vibrations of polymer chains. Pure Appl. Chem. **36**, 93–108 (1973)
21. C.W. Bunn, The crystal structure of long-chain normal paraffin hydrocarbons. The "shape" of the CH_2 group. Trans. Faraday Soc. **35**, 482–491 (1939)
22. C.W. Bunn, E.R. Howells, Structures of molecules and crystals of fluoro-carbons. Nature **174**, 549–551 (1954)
23. E. Clark, The molecular conformations of polytetrafluoroethylene: forms II and IV. Polymer **40**, 4659–4665 (1999)
24. S.-I. Mizushima, T. Simanouti, Raman frequencies of n-paraffin molecules. J. Am. Chem. Soc. **71**, 1320–1324 (1949)
25. V. LaGarde, H. Prask, S. Trevino, Vibrations in teflon. Discuss. Faraday Soc. **48**, 15–18 (1969)
26. TURBOMOLE v6.4 2012, a development of University of Karlsruhe and Forschungszentrum Karlsruhe GmbH, 1989–2007, TURBOMOLE GmbH, since 2007; available from http://www.turbomole.com
27. G. Herzberg, *Molecular Spectra and Molecular Structure: Infrared and Raman spectra of polyatomic molecules*, vol. II (R. E. Krieger Pub. Co., New York, 1991)
28. S.J. Daunt, H. Shurvell, The gas phase infrared band contours of s-triazine and s-triazine-d_3: The fundamentals of $C_3N_3H_3$ and $C_3N_3D_3$ and some overtone and combination bands of $C_3N_3H_3$. J. Mol. Spectro. **62**, 373–395 (1976)
29. J. Thelemann, Accordion Vibrations of Linear Alkanes, B.Sc. Thesis, Georg-August-Universität Göttingen, 2009
30. T.N. Wassermann, Umgebungseinflüsse auf die C-C- und C-O-Torsionsdynamik in Molekülen und Molekülaggregaten: Schwingungsspektroskopie bei tiefen Temperaturen, Ph.D. Thesis, Georg-August-Universität Göttingen, 2009
31. M. Tasumi, T. Shimanouchi, T. Miyazawa, Normal vibrations and force constants of polymethylene chain. J. Mol. Spectro. **9**, 261–287 (1962)
32. R.R. Laher, F.R. Gilmore, Improved fits for the vibrational and rotational constants of many states of nitrogen and oxygen. J. Phys. Chem. Ref. Data **20**, 685–712 (1991)
33. I.M. Grigoriev, A.V. Domanskaya, A.V. Podzorov, M.V. Tonkov, Intra- and intermolecular components of the ν_2 forbidden band of CF_4 in pure gas and in He, Ar, Xe and N_2 mixtures. Mol. Phys. **102**, 1851–1857 (2004)
34. V. Boudon, J. Mitchell, A. Domanskaya, C. Maul, R. Georges, A. Benidar, W.G. Harter, High-resolution spectroscopy and analysis of the $\nu_3/2\nu_4$ dyad of CF_4. Mol. Phys. **109**, 2273–2290 (2011)
35. P. Drawe, Mechanical properties of chain molecules from spectroscopic data, B.Sc. Thesis, Georg-August-Universität Göttingen, 2010
36. M.J. Frisch, G.W. Trucks, H.B. Schlegel, G.E. Scuseria, M.A. Robb, J.R. Cheeseman, J.A. Montgomery, Jr., T. Vreven, K.N. Kudin, J.C. Burant, J.M. Millam, S.S. Iyengar, J. Tomasi, V. Barone, B. Mennucci, M. Cossi, G. Scalmani, N. Rega, G.A. Petersson, H. Nakatsuji, M. Hada, M. Ehara, K. Toyota, R. Fukuda, J. Hasegawa, M. Ishida, T. Nakajima, Y. Honda, O.

Kitao, H. Nakai, M. Klene, X. Li, J. E. Knox, H.P. Hratchian, J.B. Cross, V. Bakken, C. Adamo, J. Jaramillo, R. Gomperts, R.E. Stratmann, O. Yazyev, A.J. Austin, R. Cammi, C. Pomelli, J.W. Ochterski, P.Y. Ayala, K. Morokuma, G.A. Voth, P. Salvador, J.J. Dannenberg, V.G. Zakrzewski, S. Dapprich, A.D. Daniels, M.C. Strain, O. Farkas, D. K. Malick, A.D. Rabuck, K. Raghavachari, J.B. Foresman, J.V. Ortiz, Q. Cui, A.G. Baboul, S. Clifford, J. Cioslowski, B.B. Stefanov, G. Liu, A. Liashenko, P. Piskorz, I. Komaromi, R.L. Martin, D.J. Fox, T. Keith, M. A. Al-Laham, C.Y. Peng, A. Nanayakkara, M. Challacombe, P.M.W. Gill, B. Johnson, W. Chen, M.W. Wong, C. Gonzalez, J.A. Pople, Gaussian 03, Revision B.04, Gaussian Inc, Wallingford, CT (2004)

37. T. Shimanouchi, in *NIST Chemistry WebBook, NIST Standard Reference Database Number 69*, ed. by P.J. Linstrom, W.G. Mallard (National Institute of Standards and Technology, Gaithersburg MD, 20899, 2013), Chap. Molecular Vibrational Frequencies
38. I. Sakurada, K. Kaji, S. Wadano, Elastic moduli and structure of low density polyethylene. Colloid Polym. Sci. **259**, 1208–1213 (1981)
39. P.J. Barham, A. Keller, High-strength polyethylene fibres from solution and gel spinning. J. Mater. Sci. **20**, 2281–2302 (1985)
40. A. Peterlin, Drawing and extrusion of semi-crystalline polymers. Colloid Polym. Sci. **265**, 357–382 (1987)
41. B. Fanconi, J.F. Rabolt, The determination of longitudinal crystal moduli in polymers by spectroscopic methods. J. Polym. Sci. Polym. Phys. Ed. **23**, 1201–1215 (1985)

Chapter 7
Summary and Outlook

Atticus told me to delete the adjectives and I'd have the facts.
Harper Lee

The focus of the present thesis was the identification and quantification of dispersion-induced self-solvation (folding) of unbranched n-alkanes, free of interactions with a molecular environment. To this end, a sensitive Raman jet-spectrometer denoted curry-jet was extended to allow preparation of low-volatile substances, and n-alkanes as long as 21 carbon atoms were isolated in supersonic jet-expansions for the first time. Vibrational Raman spectra of structure-dependent frame vibrations and structurally less dependent C–C and C–H stretching vibrations were recorded and systematically analyzed. Quantum chemical simulations, using the B3LYP method augmented with Grimme's dispersion correction from 2010 [1], were made and compared to experimental spectra to aid the assignment and estimate the abundances of different conformers. The highlights of the findings are as follows.

- The most important contribution to Raman jet-spectra of frame vibrations comes from single gauche and all-trans conformers up to a chain length of 18 carbon atoms, where first evidence of self-solvation is found. Beyond this chain length, bands from single gauche conformers fade into the background, and bands from self-solvating "hairpin" conformers steeply gain intensity with increasing chain length ($\approx$150–300 cm^{-1}, spectra on page 87).
- It is proposed that the onset of hairpin formation at a chain length of 18 carbon atoms marks the critical chain length n_c beyond which hairpin conformers are more stable than all-trans conformers ($n_c = 17$ or 18), despite the fact that the hairpin abundance does not outweigh the all-trans abundance up until a chain length of $n = 21$, which is explained by a kinetic delay of self-solvation in jet-expansions (discussion on page 107).
- In agreement with the assessment of the Raman jet-experiments, high level coupled cluster calculations find $n_c = 17 \pm 1$ [2] (discussion on page 110).

N.O.B. Lüttschwager, *Raman Spectroscopy of Conformational Rearrangements at Low Temperatures*, Springer Theses, DOI 10.1007/978-3-319-08566-1_7

The spectral evidence for folding is very abundant and not restricted to the region of low-frequency frame vibrations (Sect. 4.4.4). Strongly Raman-scattering frame vibrations of hairpin conformers are accordion-like modes with wavenumbers correlated to the length of vibrating all-trans segments (discussion on page 80). In the region of C–C stretching vibrations, one finds hairpin vibrations localized in the kink of the hairpin conformer (887 cm^{-1}) or such shifted from single gauche and all-trans in-phase C–C stretching vibrations ($\approx$1110 and 1145 cm^{-1}, spectra on page 90). Asymmetric broadening of the CH_2 stretching band at 2860 cm^{-1} and rather symmetric broadening of the CH_3 stretching band at 2870 cm^{-1}, setting in at a chain length of 17 carbon atoms, are further features which may be related to self-solvation (figures on page 98 and 102).

Quantum chemical simulations (B3LYP-D3/def2-TZVP, Sect. 4.3) were used intensely to interpret jet-spectra and estimate conformer fractions. The excellent performance of dispersion corrected B3LYP regarding harmonic frequencies and conformer energies is worth mentioning in this summary. The correct energy prediction depends critically on the inclusion of the dispersion correction "D3" [1] (Appendix A.1). The critical chain length on the B3LYP-D3/def2-TZVP level is $n_c = 16$, close to the coupled cluster result and onset of hairpin formation in supersonic expansions.

Perfluorinated alkanes in the chain length interval $n = 6$–14 were studied by Raman jet-spectroscopy as a comparative case. The main goal was the extrapolation of the elastic modulus of polytetrafluoroethylene from accordion vibration wavenumbers. In the same way, the elastic modulus of polyethylene was determined. The major findings of these studies are:

- Other than in case of n-alkanes, the population of gauche conformers is negligible for jet-isolated perfluorinated alkanes (reflected in 'mono-conformational' spectra, page 121), which points to a higher gauche energy and/or smaller isomerization barrier.
- Self-solvation does not play a role in case of perfluorinated alkanes in the considered size range, and it will most probably not play a role up until a chain length much longer than the critical chain length of n-alkanes (discussion on page 122).
- The ultimate elastic modulus of cold, interaction-free polyethylene is (305 ± 5) GPa, the one of polytetrafluoroethylene (209 ± 10) GPa. The forces necessary to stretch an all-trans polyethylene chain or a slightly helical all-anti polytetrafluoroetyhlene chain are virtually identical. The different elastic moduli are merely a result of different chain packing (discussion on page 138), as noted in an earlier study [3].

Other than that, it could be shown that the application of low-level quantum chemical calculations to aid the extrapolation of the elastic modulus of polyethylene from accordion vibration frequencies [4] is not critical, which in return supports the extrapolation of the elastic modulus of polytetrafluoroethylene using low-level quantum chemistry, performed here.

Regarding the extrapolation of elastic moduli, it should be relatively straightforward to prepare perfluorinated alkanes longer than perfluorotetradecane in jet-expansions with the heated version of the curry-jet, but the accordion vibrations in this size range would reach the cutoff-wavelength of the edge-filter used to suppress Rayleigh-scattered light. A filter which allows measurements closer to the Rayleigh-line would thus be needed to measure accordion vibrations of long perfluorinated alkanes. Also, rotational lines from air impurities overlay in the region close to the Rayleigh-line, since the heatable brass saturator used to prepare substances above room temperature is somewhat leaking. This problem might be significantly reduced in a revised version, which is currently being tested.

The study of *n*-alkane self-solvation with the curry-jet calls for some further experiments. It would be very interesting to probe the influence of the nozzle temperature on the abundance of hairpin conformers. For measurements of octadecane, the nozzle temperature may be lowered from 130 °C to about 90–100 °C. Depending on how much of the hairpin jet-abundance stems from relaxation and how much comes from freezing the nozzle-temperature distribution, a lower nozzle temperature might significantly increase the overall hairpin abundance, which would raise the question if it is possible to prepare and observe hairpin conformers of heptadecane or even shorter alkanes without competition from clustering. If so, this would challenge the critical chain length assigned here. It would also be helpful to employ an optical resonator to enhance the intensity of the excitation radiation [5]. In this way, it would be possible to work with lower alkane substance temperatures and thus even lower nozzle temperatures. The construction of a 'power build-up cavity' [6] for the curry-jet was initiated during the work for this dissertation and will be continued in future doctoral work. With the curry-jet in its current configuration, one could also study partially or completely deuterated alkanes to probe the influence of zero-point vibrational energy or to isolate specific vibrations [7–9].

An effort which was not mentioned in this thesis was a Monte-Carlo simulation made with Matlab [10] which models the conformational distribution of *n*-alkanes during a temperature drop to assess the kinetic delay thought to reduce the hairpin abundance in jet-expansions. This simulation is in agreement with the expectation that the hairpin conformer takes substantially more isomerization steps to be formed in high abundance compared to the all-trans conformer, which is formed quickly. However, the determination of conformational energies, especially from conformers similar to the hairpin conformer, is not yet implemented satisfactorily. This is critical because it will affect the rate with which hairpin conformers are formed when the temperature is lowered. Therefore, the discussion of this simulation was omitted. More sophisticated simulations [11] of a temperature drop could help to quantify the kinetic delay.

Most desirable would be experiments which reach even lower effective conformational temperatures than those reached in this work [12, 13], with conformational distributions closer to thermal equilibrium. To this end, it would be necessary to cool more slowly, providing more time for hairpin conformers to equilibrate, but at the same time prevent condensation and keeping the molecules isolated. A promising candidate might be buffer gas cooling [14, 15], that is, passing alkane vapor into helium gas kept at $\approx$6 K which provides collisional cooling. With this technique, the preparation of relatively large molecules at very low-temperatures has already been demonstrated [16]. As in case of supersonic expansions, however, the number of collisions is limited, because the substance is eventually lost on cold cell walls. It is an open question if conformational degrees of freedom of alkanes exposed to a cold buffer gas would equilibrate sufficiently fast to provide a more favorable conformer distribution than supersonic expansions.

References

1. S. Grimme, J. Antony, S. Ehrlich, H. Krieg, A consistent and accurate ab initio parametrization of density functional dispersion correction (DFT-D) for the 94 elements H-Pu. J. Chem. Phys. **132**, 154104 (2010)
2. N.O.B. Lüttschwager, T.N. Wassermann, R.A. Mata, M.A. Suhm, The last globally stable extended alkane. Angew. Chem. Int. Ed. **52**, 463–466 (2013)
3. I. Sakurada, K. Kaji, Relation between the polymer conformation and the elastic modulus of the crystalline region of polymer. J. Polym. Sci. Part C: Polym. Symp. **31**, 57–76 (1970)
4. T.N. Wassermann, J. Thelemann, P. Zielke, M.A. Suhm, The stiffness of a fully stretched polyethylene chain: a Raman jet spectroscopy extrapolation. J. Chem. Phys. **131**, 161108 (2009)
5. R. Salter, J. Chu, M. Hippler, Cavity-enhanced Raman spectroscopy with optical feedback cw diode lasers for gas phase analysis and spectroscopy. Analyst **137**, 4669–4676 (2012)
6. J.A. Barnes, T.E. Gough, M. Stoer, Laser power build-up cavity for high-resolution laser spectroscopy. Rev. Sci. Instrum. **70**, 3515–3518 (1999)
7. R.A. MacPhail, R.G. Snyder, H.L. Strauss, The motional collapse of the methyl C–H stretching vibration bands. J. Chem. Phys. **77**, 1118–1137 (1982)
8. R.G. Snyder, A.L. Aljibury, H.L. Strauss, H.L. Casal, K.M. Gough, W.F. Murphy, Isolated C–H stretching vibrations of *n*-alkanes: assignments and relation to structure. J. Chem. Phys. **81**, 5352–5361 (1984)
9. Z. Liao, J.E. Pemberton, Raman spectral conformational order indicators in perdeuterated alkyl chain systems. J. Phys. Chem. A **110**, 13744–13753 (2006)
10. MATLAB, version 7.12.0 (R2011a). The MathWorks Inc., Natick (2011)
11. L.L. Thomas, T.J. Christakis, W.L. Jorgensen, Conformation of alkanes in the gas phase and pure liquides. J. Phys. Chem. B **110**, 21198–21204 (2006)
12. M. Schnell, G. Meijer, Cold molecules: preparation, applications, and challenges. Angew. Chem. Int. Ed. **48**, 6010–6031 (2009)
13. M. Schnell, G. Meijer, Kalte Moleküle: Herstellung, Anwendungen und Herausforderungen. Angew. Chem. **121**, 6124–6147 (2009)
14. N.R. Hutzler, H.-I. Lu, J.M. Doyle, The buffer gas beam: an intense, cold, and slow source for atoms and molecules. Chem. Rev. **112**, 4803–4827 (2012)
15. D. Patterson, J.M. Doyle, Cooling molecules in a cell for FTMW spectroscopy. Mol. Phys. **110**, 1757–1766 (2012)
16. D. Patterson, E. Tsikata, J.M. Doyle, Cooling and collisions of large gas phase molecules. Phys. Chem. Chem. Phys. **12**, 9736–9741 (2010)

Appendix

A.1 Gauche Energy and Additivity Check

Electronic energies of (perfluoro-)hexane conformers calculated with Turbomole v6.4 [1] are visualized in Figs. A.1, A.2 and A.3. The calculations were done as described in Sect. 4.3.[1] MP2 energies are included for comparison, calculated employing the resolution-of-identity (RI) approximation to reduce the computational cost, with the Turbomole program ricc2 [2]. For *n*-hexane, the best estimates from Gruzman et al. [3] (coupled cluster extrapolations) are included as reference.

The bar plots show calculated electronic energies (from center to right) and corresponding added single gauche energies (from center to left) to evaluate gauche energy additivity [4]. Narrower bars drawn with lighter colors indicate the deviation from simple additivity to the "true" conformer energy (at that particular level of theory). For example: the ttg and tgt hexane conformers have electronic energies 2.55 and 2.64 kJ mol^{-1} higher than the electronic energy of the all-trans conformer on the B3LYP-D3/def2-TZVP level, respectively. Assuming perfect additivity and adding these energies leads to a hypothetical energy of 5.19 kJ mol^{-1} for the tgg conformer. However, the calculated energy is 4.28, 0.91 kJ mol^{-1} less than the added energies. This demonstrates the stabilizing "positive pentane effect" introduced in Sect. 4.1, which was attributed to beneficial dispersion interactions [4].

It is interesting to extend this comparison to other conformers, levels of theory, and especially to perfluorinated alkanes. Calculations of hexane conformers on different levels of theory are summarized in Fig. A.1. As expected, one finds a stabilization of conformers with adjacent gauche bonds of equal rotational sense, and a large destabilization if the rotational sense is reversed. Additivity is perfectly realized in case two gauche bonds of equal rotational sense are separated by a trans bond. Two trans-separated gauche bonds of opposite rotational sense are slightly destabilized. The different computational levels compare favorably with exception of non-dispersion-corrected B3LYP calculations. In this case, single gauche energies are erroneously

[1] Energy convergence criterion $\leq 10^{-7}$ hartree, gradient norm convergence criterion $\leq 10^{-4}$ hartree $bohr^{-1}$. B3LYP calculations were done on the m4 grid.

N.O.B. Lüttschwager, *Raman Spectroscopy of Conformational Rearrangements at Low Temperatures*, Springer Theses, DOI 10.1007/978-3-319-08566-1

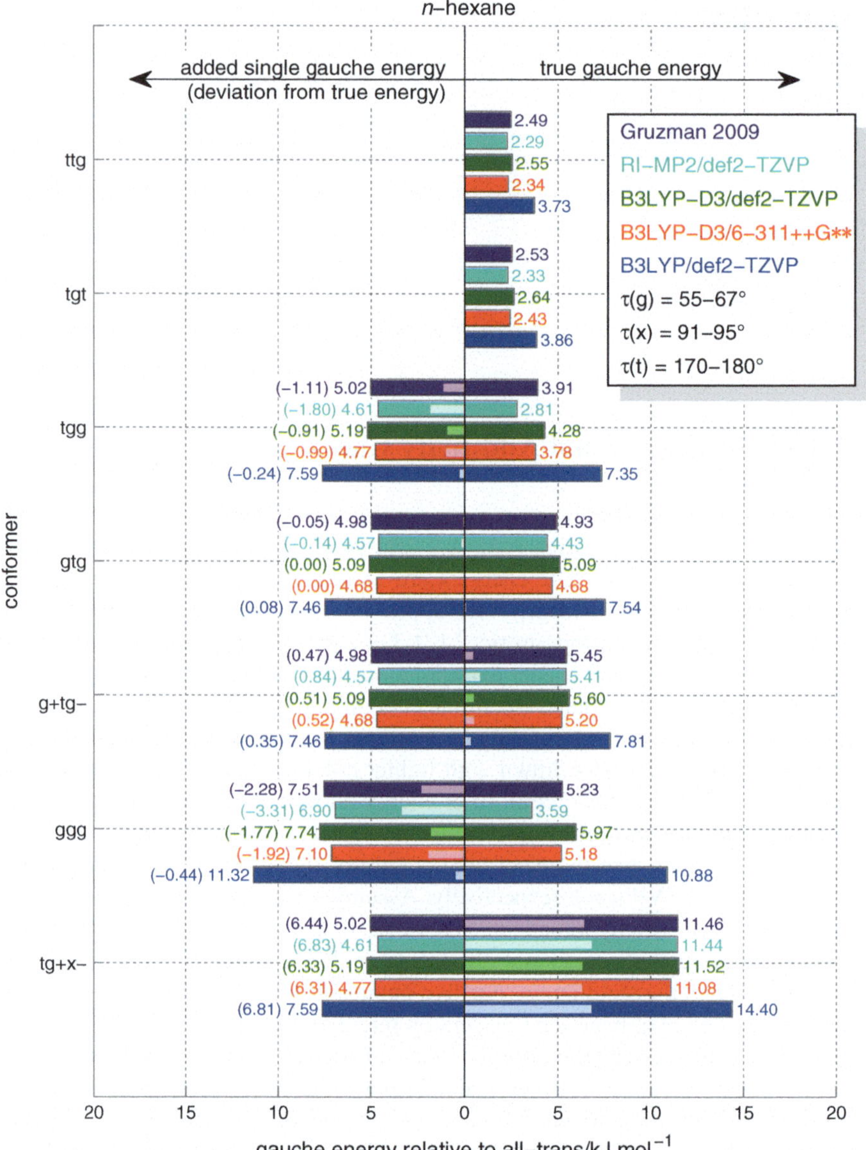

Fig. A.1 Electronic energies of *n*-hexane conformers calculated with Turbomole v6.4 [1]. A coupled cluster extrapolation from Gruzman et al. [3] is taken as reference. τ denotes the torsion angles of gauche (g), cross (x), and trans (t) conformations. In case no sign is given, torsional angles are of equal rotational sense. The light bars show deviation from additivity (from *center* to the *left* stabilizing, from *center* to the *right* destabilizing)

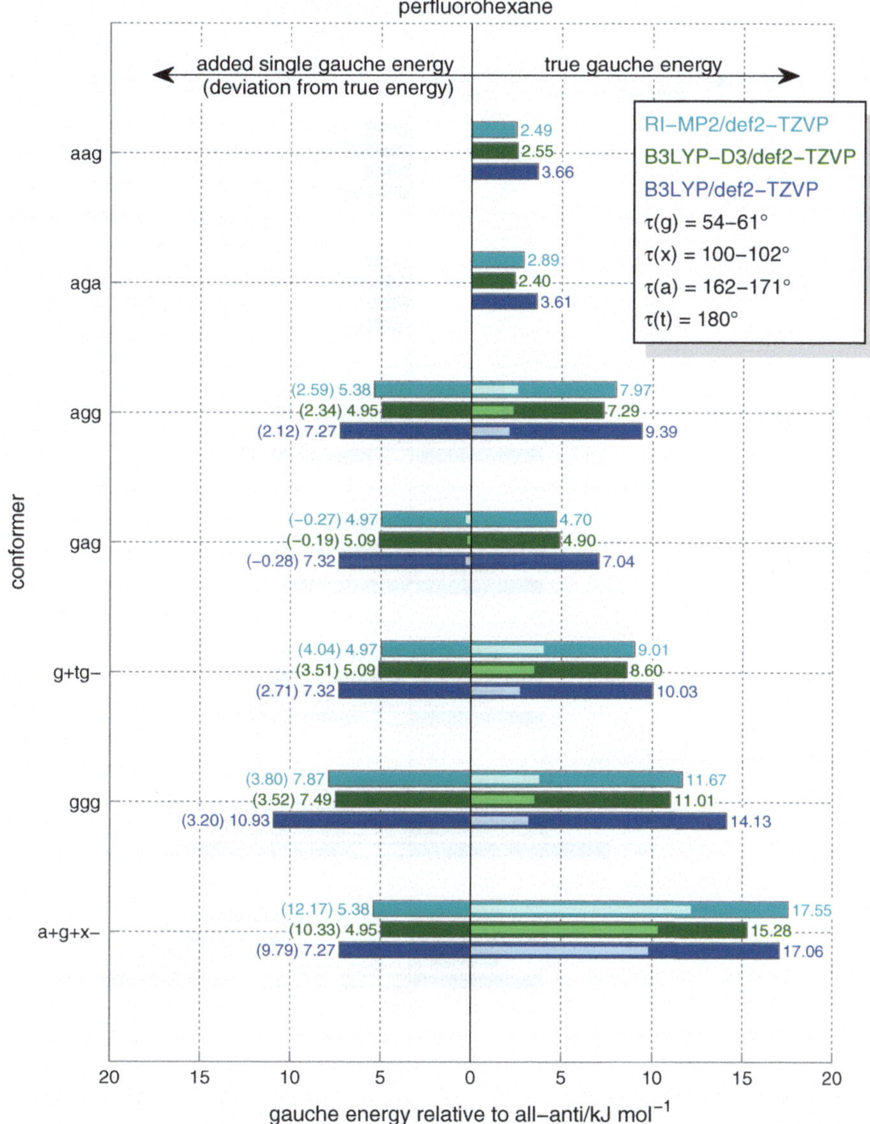

Fig. A.2 Electronic energies of perfluorohexane conformers calculated with Turbomole v6.4 [1]. τ denotes the torsion angles of gauche (g), cross (x), anti (a), and trans (t) conformations. In case no sign is given, torsional angles are of equal rotational sense. The light bars show deviation from additivity (from *center* to the *left* stabilizing, from *center* to the *right* destabilizing)

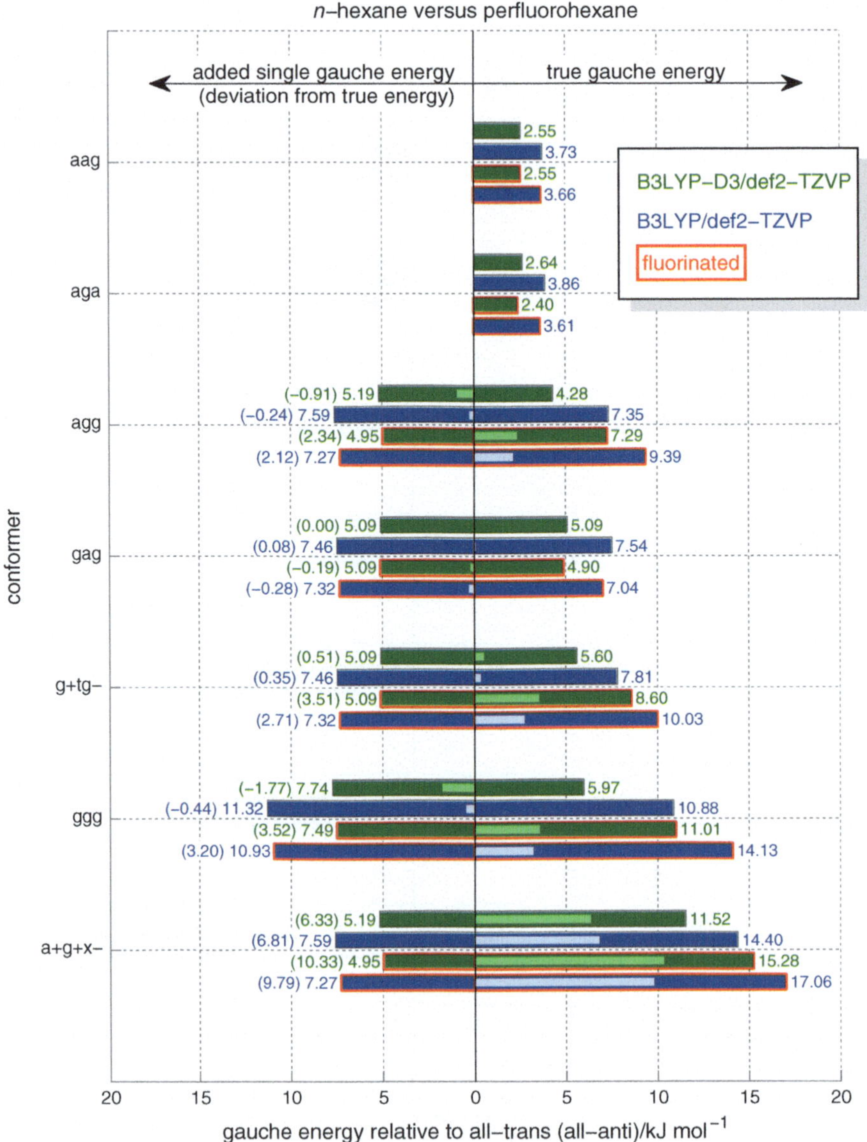

Fig. A.3 Comparison of relative electronic energies of *n*-hexane and perfluorohexane conformers calculated with Turbomole v6.4 [1]. τ denotes the torsion angles of gauche (g), cross (x), anti (a), and trans (t) conformations. In case no sign is given, torsional angles are of equal rotational sense. The *light bars* show deviation from additivity (from *center* to the *left* stabilizing, from *center* to the *right* destabilizing)

high and the dispersion stabilization due to neighboring gauche bonds is largely missing. MP2 calculations, on the other hand, overestimate dispersion stabilization [5–7], which is reflected in too small energies of conformers involving a $g^{\pm}g^{\pm}$ sequence.

For perfluorinated alkanes, B3LYP calculations with and without dispersion correction [8] as well as MP2 calculations come to very similar results regarding additivity. Other than in case of *n*-hexane, two adjacent gauche bonds of equal rotational sense lead to a conformer energy higher than the sum of single gauche energies, implying a destabilization. If two gauche bonds of equal rotational sense are separated by a bond in anti conformation, the energy is additive, but if the gauche bonds are twisted in opposing directions, the bond separating them is forced from the preferred helical anti conformation to a true trans conformation ($\tau = 180°$) and the conformer energy increases notably. The higher steric demands of fluorine atoms thus seem to largely inhibit folding of perfluorinated alkanes of any kind, and to prevent simple additivity when gauche conformations are close to each other.

The experimental energy of a perfluorinated single gauche conformer remains elusive. From the experimental viewpoint, it should be considerably higher than in case of normal alkanes (see Sect. 5), but this is at variance with the MP2 and dispersion corrected B3LYP predictions presented here. Predictions of bare B3LYP are in better agreement with the qualitative experimental observations, but overshoot recent coupled cluster calculations which found a smaller gauche energy of 3.1 kJ mol^{-1} for perfluorobutane [9] on the CCSD/cc-pVTZ level.

A.2 Fortran 95 Program `count_conf.f95`

The program `count_conf` starts by calculating the maximum number of C–C–C–C torsions (`max_torsions`) from the user defined chain length and allocates memory accordingly. It enters a `do` loop which runs from 1 to the number of branches in the tree diagram. The path along each branch is uniquely defined by the indexing number of the branch stored in the variable `branch`. Dividing this integer number by 3 gives a smaller number with the fractional digits of either 0, 1/3 or 2/3 (variable `rest`). The first element of the branch is selected according to the fractional digits of the division `branch/3.0` ($0 \rightarrow -1$ for g^-, $1/3 \rightarrow 0$ for t and $2/3 \rightarrow +1$ for g^+). Subsequent divisions of the rounded resulting integers yield the remaining elements. The selection is done with the `select case` statement (line 21) which needs an integer selector (`nint(rest*3)`). Each cycle of the `do` loop yields a torsional sequence temporarily stored in the array `conf`, for example `conf = [0 0 +1 0 -1]` for the $n = 8$ conformer ttg^+tg^-. The number of gauche torsions is calculated by adding the absolute values of the elements in `conf`. The corresponding element in the array `coeff` is readily raised using the number of gauche bonds for indexing (line 37–39). All-trans is not counted, as its coefficient is known and the sum of the elements would be 0 which is not a valid index. Rejection of $g^{\pm}g^{\mp}$ sequences is done by comparing the current torsion with the preceding torsion (`last_torsion`) in the current branch (line 31–33). Before the code displays the resulting coefficient on the screen, the numbers are divided by 2 to include symmetry weighting. The code is not optimized for speed, as the calculations for even the longest alkane considered in this work ($n = 21$ with $3^{18} \approx 3.9 \times 10^8$ main loops/branches and $18 \times 3^{18} \approx 7.0 \times 10^9$ inner loops/elements) take no more than a few minutes on an up-to-date desktop PC.

```
1  program count_conf
2  implicit none
3  double precision division, rest
4  integer branch, chain_length, i, last_torsion,
   max_torsions, torsion
5  integer, dimension(:), allocatable :: conf, coeff
6  ! get chain length from user input
7  print *,'Chain length:'
8  read *, chain_length
9  max_torsions = chain_length - 3
10 allocate(conf(max_torsions))
11 allocate(coeff(max_torsions))
12 coeff(1:max_torsions) = 0
13 ! main loop to go through each branch of the tree diagram
14 do branch = 1, (3**max_torsions)
15   conf(1:max_torsions) = 0
16   last_torsion = 0
17   division = dble(branch) / 3.0
18   ! inner loop to go through each element
19   do torsion = 1 , max_torsions
20     rest = division - aint(division)
21     select case (nint(rest*3))
22       case (0)
23       conf(torsion) = -1
24       case (1)
25         conf(torsion) = 0
26         division = aint(division) + 1.0
27       case (2)
28         conf(torsion) = 1
29         division = aint(division) + 1.0
30     end select
31     if (last_torsion /= 0 .and. conf(torsion) ==
       -last_torsion) then
32       goto 100
33     end if
34     last_torsion = conf(torsion)
35     division = division / 3.0
36     end do
```

```
    if (sum(abs(conf)) /= 0) then
      coeff(sum(abs(conf))) = coeff(sum(abs(conf))) + 1
    end if
    100 cycle
  end do
  ! output of number of gauche vs. coefficient
  print *,'coefficient for i gauche:'
  do i = 1, max_torsions
    print *, i,'gauche:', coeff(i) / 2
  end do
  end program
```

A.3 Gaussian Functions to Simulate Raman Spectra

For the simulation of Raman spectra, each vibrational band (index v) is modeled by a Gaussian function of the form

$$f_v(\tilde{\nu}) = \sigma'_v \cdot \exp\left(-\frac{\ln(16) \cdot (\tilde{\nu} - \tilde{\nu}_v)^2}{\mathrm{FWHM}^2}\right),$$

where σ'_v is the Raman scattering cross section and $\tilde{\nu}_v$ the wavenumber of the specific vibration. The factor $\ln(16)$ ensures that

$$f_v(\tilde{\nu}_v \pm \mathrm{FWHM}/2) = \sigma'_v/2,$$

so that FWHM is indeed the full width at half maximum. The integral over the whole wavenumber interval is

$$\int_{-\infty}^{\infty} f_v(\tilde{\nu})\mathrm{d}\tilde{\nu} = \sqrt{\pi/\ln(16)} \cdot \sigma'_v \cdot \mathrm{FWHM},$$

directly proportional to the scattering cross section and the FWHM. The spectrum of a single conformer (index j) is obtained by summing over all vibrational bands:

$$f_j(\tilde{\nu}) = \sum_v f_v(\tilde{\nu}),$$

and the complete simulated spectrum by summing over all conformers:

$$F(\tilde{\nu}) = \sum_j \frac{N_j}{N_i} f_j(\tilde{\nu}).$$

In the last equation, (N_j/N_i) is a Boltzmann weighting factor, discussed in Sect. 4.1.

A.4 Gibbs Energy Difference from Quantum Chemical Calculations

The following derivation makes use of equations from thermodynamics and statistical mechanics taken from Ref. [10]. For simplicity, all equations will be derived for an ensemble containing 6.022141×10^{23} (1 mol) molecules, equivalent with replacing the number of considered molecules (N) by Avogadro's constant (N_{A}) in the following equations.

The internal energy (U) of an ensemble of molecules is given by the electronic energy (E_{SCF}), obtained from self-consistent field (SCF) calculations, zero-point vibrational energy (E_{ZPVE}), obtained from harmonic frequency calculations, and energy from thermal excitation of the various degrees of freedom of the system ($E_{\mathrm{th}}(T)$):

$$U = E_{\mathrm{SCF}} + E_{\mathrm{ZPVE}} + E_{\mathrm{th}}(T).$$

E_{SCF} and E_{ZPVE} are provided by Turbomole. The thermal energy is connected to the canonical partition function (Q) by:

$$E_{\mathrm{th}}(T) = k_{\mathrm{B}} T^2 \left(\frac{\partial \ln Q}{\partial T} \right)_{N=N_{\mathrm{A}}, V.} \tag{A.1}$$

The enthalpy includes the product of the pressure (p) and Volume (V) of the system:

$$H = U + pV,$$
$$H = E_{\mathrm{SCF}} + E_{\mathrm{ZPVE}} + \underbrace{k_{\mathrm{B}} T^2 \left(\frac{\partial \ln Q}{\partial T} \right)_{N=N_{\mathrm{A}}, V}}_{=E_{\mathrm{th}}(T)} + V \underbrace{k_{\mathrm{B}} T \left(\frac{\partial \ln Q}{\partial V} \right)_{N=N_{\mathrm{A}}, T.}}_{=p} \tag{A.2}$$

To calculate the Gibbs energy, the enthalpy is reduced by the entropy (S) times the temperature:

$$G = H - TS. \tag{A.3}$$

The entropy term is related to the canonical partition function by:

$$TS = \underbrace{k_{\mathrm{B}} T^2 \left(\frac{\partial \ln Q}{\partial T} \right)_{N=N_{\mathrm{A}}, V}}_{=E_{\mathrm{th}}(T)} + k_{\mathrm{B}} T \ln(Q). \tag{A.4}$$

Combining Eqs. A.4, A.3, A.2, and A.1 gives:

$$G = E_{\mathrm{SCF}} + E_{\mathrm{ZPVE}} + V k_{\mathrm{B}} T \left(\frac{\partial \ln Q}{\partial V} \right)_{N=N_{\mathrm{A}}, T} - k_{\mathrm{B}} T \ln(Q). \tag{A.5}$$

For non-interacting, indistinguishable molecules, the canonical partition function can be linked to the molecular partition function (q) by:

$$Q(N, V, T) = \frac{[q(V, T)]^N}{N!}.$$

Furthermore, if the degrees of freedom of a molecule described by q do not interact (for example, the rotational states are not influenced by vibrational excitation), q can be written as the product

$$q(V, T) = q_{\mathrm{el}}(T) q_{\mathrm{trans}}(V, T) q_{\mathrm{rot}}(T) q_{\mathrm{vib}}(T).$$

The dependences on T, V and N are omitted for better readability in the following equations. The quantity important to Eq. A.5, $\ln Q$, becomes:

$$\ln Q = N[\ln q_{\mathrm{el}} + \ln q_{\mathrm{trans}} + \ln q_{\mathrm{rot}} + \ln q_{\mathrm{vib}}] - \ln(N!). \quad \text{(A.6)}$$

Plugging Eq. A.6 into Eq. A.5, calculating for 1 mol of molecules ($N = N_{\mathrm{A}}$, $k_{\mathrm{B}} N_{\mathrm{A}} = R$), and noticing that only the translational molecular partition function q_{trans} contributes to the derivative with respect to V gives:

$$\begin{aligned} G = {} & E_{\mathrm{SCF}} + E_{\mathrm{ZPVE}} + VRT \left(\frac{\partial \ln q_{\mathrm{trans}}}{\partial V} \right)_{N=N_{\mathrm{A}},T} \\ & - RT[\ln q_{\mathrm{el}} + \ln q_{\mathrm{trans}} + \ln q_{\mathrm{rot}} + \ln q_{\mathrm{vib}}] \\ & - k_{\mathrm{B}} T \ln(N_{\mathrm{A}}!). \end{aligned} \quad \text{(A.7)}$$

For two conformers of the same alkane (index i, j), the translational molecular partition functions do not differ, so that for the Gibbs energy difference ($\Delta G_{ij} = G_j - G_i$), all terms depending on q_{trans} cancel. The last term of Eq. A.7 cancels as well if two ensembles consisting of N_{A} molecules each are compared. The electronic molecular partition function will approach 1 for temperatures relevant to this work, therefore $\ln q_{\mathrm{el}} = 0$. Then, only the rotational and vibrational parts remain relevant for the calculation of ΔG_{ij}:

$$\Delta G_{ij} = \Delta E_{ij}^0 - RT(\ln q_{j,\mathrm{rot}} - \ln q_{i,\mathrm{rot}} + \ln q_{j,\mathrm{vib}} - \ln q_{i,\mathrm{vib}})$$

$$\Delta G_{ij} = \Delta E_{ij}^0 - RT \ln \left(\frac{q_{j,\mathrm{rot}} q_{j,\mathrm{vib}}}{q_{i,\mathrm{rot}} q_{i,\mathrm{vib}}} \right) \quad \text{(A.8)}$$

In these equations, the electronic SCF and zero-point vibrational energy is summarized in E_i^0:

$$E_i^0 = E_{i,\mathrm{SCF}} + E_{i,\mathrm{ZPVE}}.$$

Finally, to account for degeneracy of conformers, the Gibbs energy difference has to be corrected by a term ($-RT \ln \frac{g_j}{g_i}$) [10, p. 563], where the gs are the numbers of enantiomeric realizations of otherwise equal conformers. This is equivalent to multiplying the molecular partition functions with $g_{i,j}$:

$$\Delta G_{ij} = \Delta E_{ij}^{0} - RT \ln \left(\frac{g_j \, q_{j,\mathrm{rot}} \, q_{j,\mathrm{vib}}}{g_i \, q_{i,\mathrm{rot}} \, q_{i,\mathrm{vib}}} \right). \tag{A.9}$$

A.5 Hairpin Normal Coordinates

In the following figures (Figs. A.4–A.10), normal coordinates of characteristic low-frequency hairpin vibrations are shown, framed by exaggerated turning points. Labels refer to the chain length, hairpin type, and experimental and unscaled calculated wavenumbers. The latter is given in square brackets. The vibrations are discussed in Sect. 4.4.1.

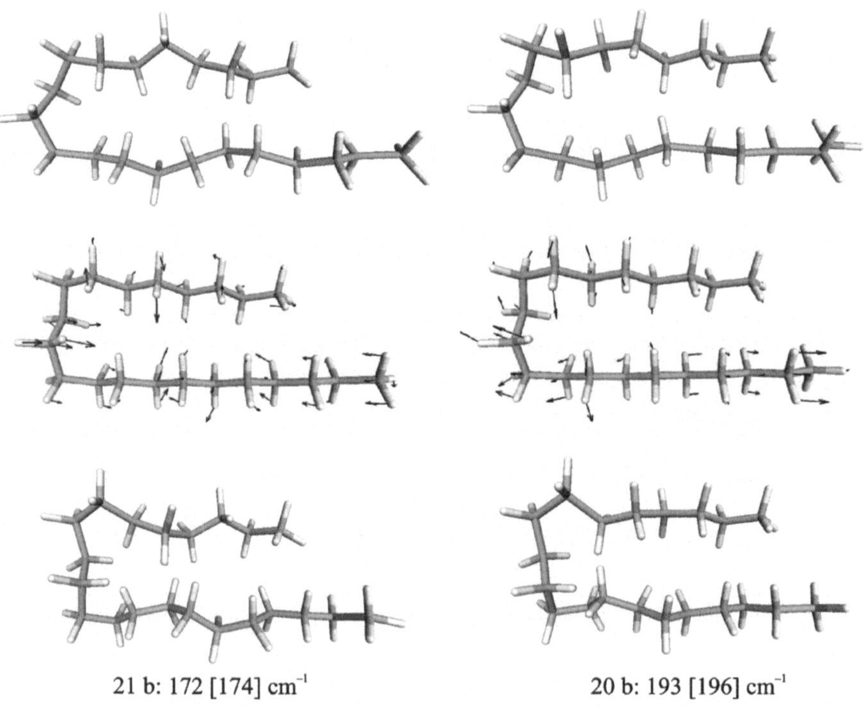

Fig. A.4 Skeletal hairpin vibrations of type I

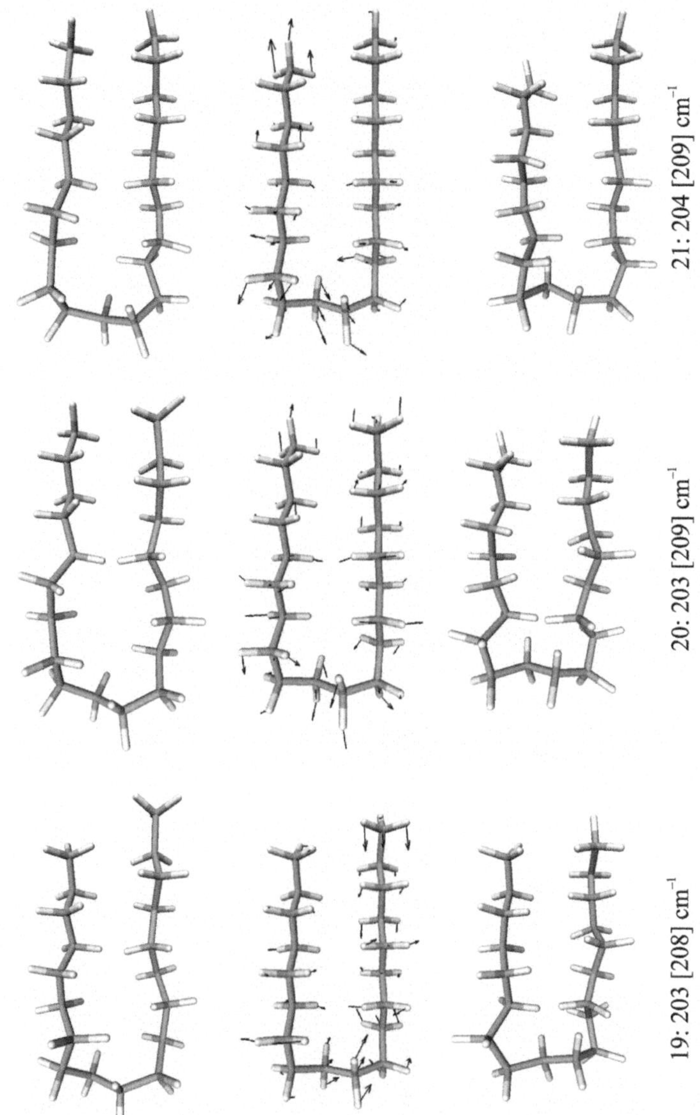

Fig. A.5 Skeletal hairpin vibrations of type I

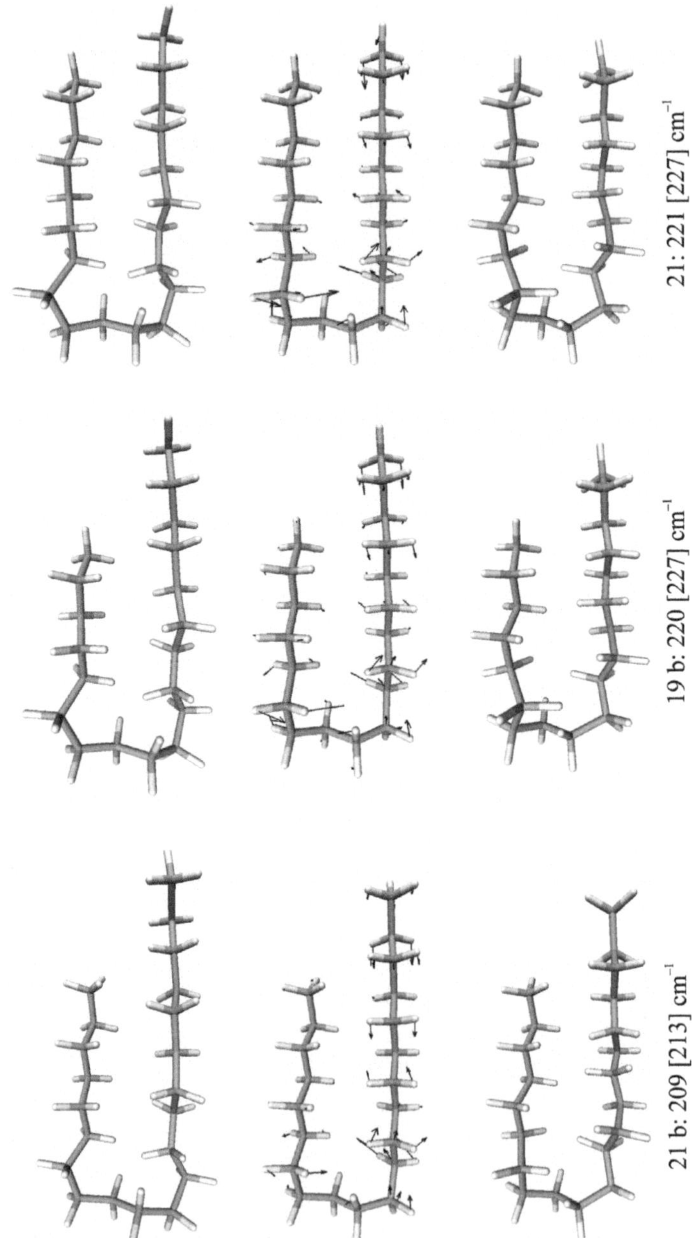

Fig. A.6 Skeletal hairpin vibrations of type II

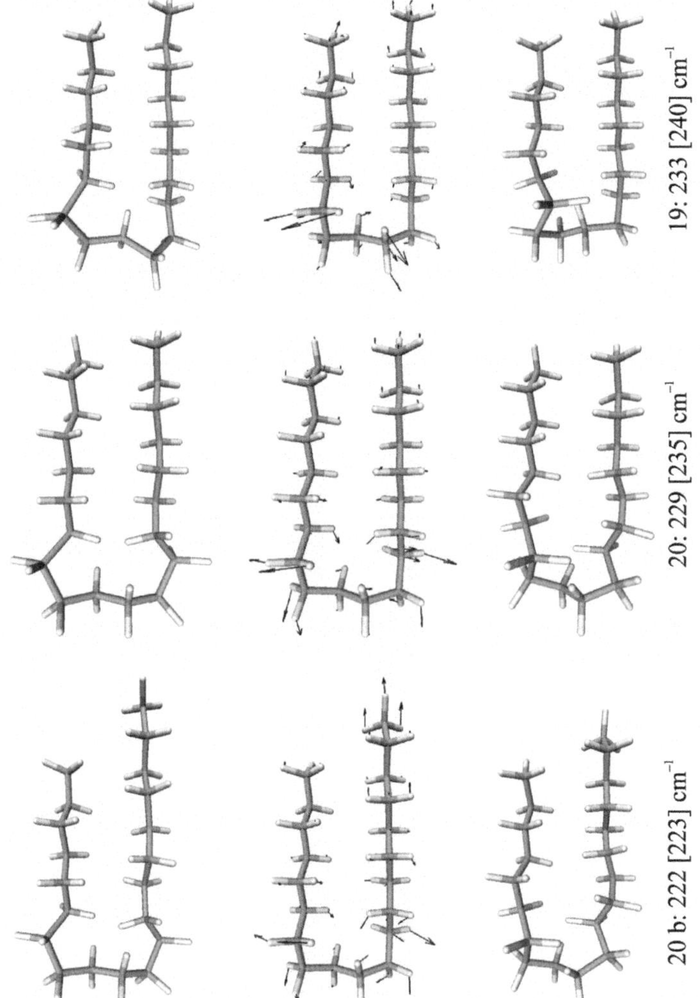

Fig. A.7 Skeletal hairpin vibrations of type II

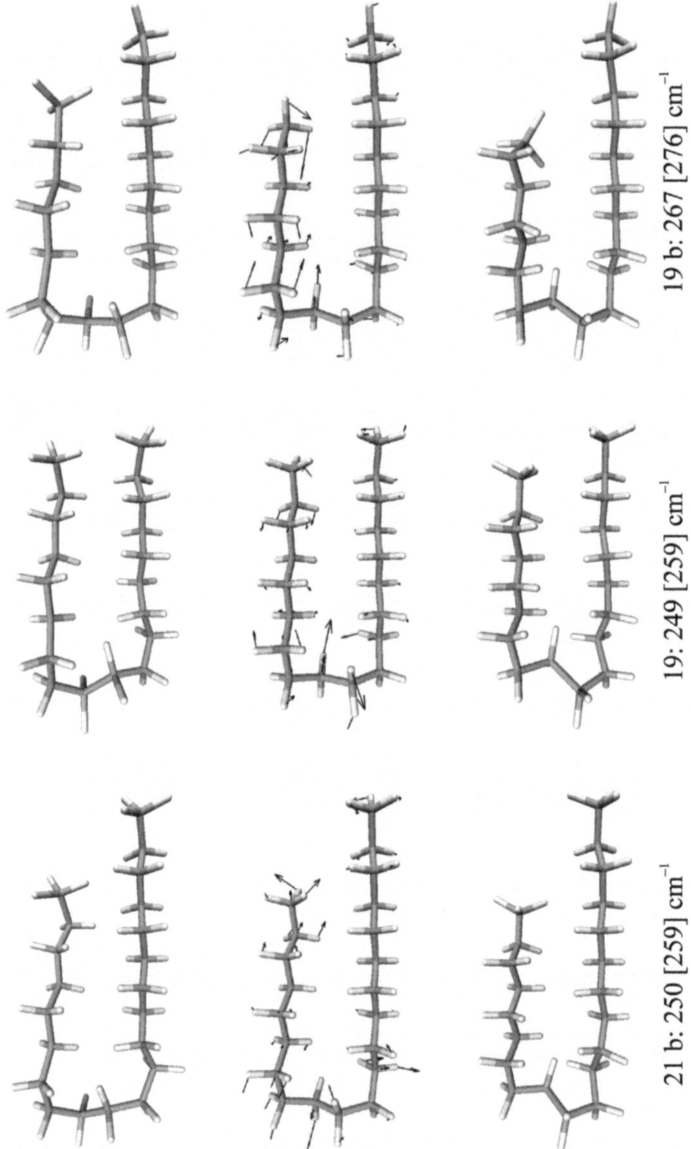

Fig. A.8 Skeletal hairpin vibrations of type III

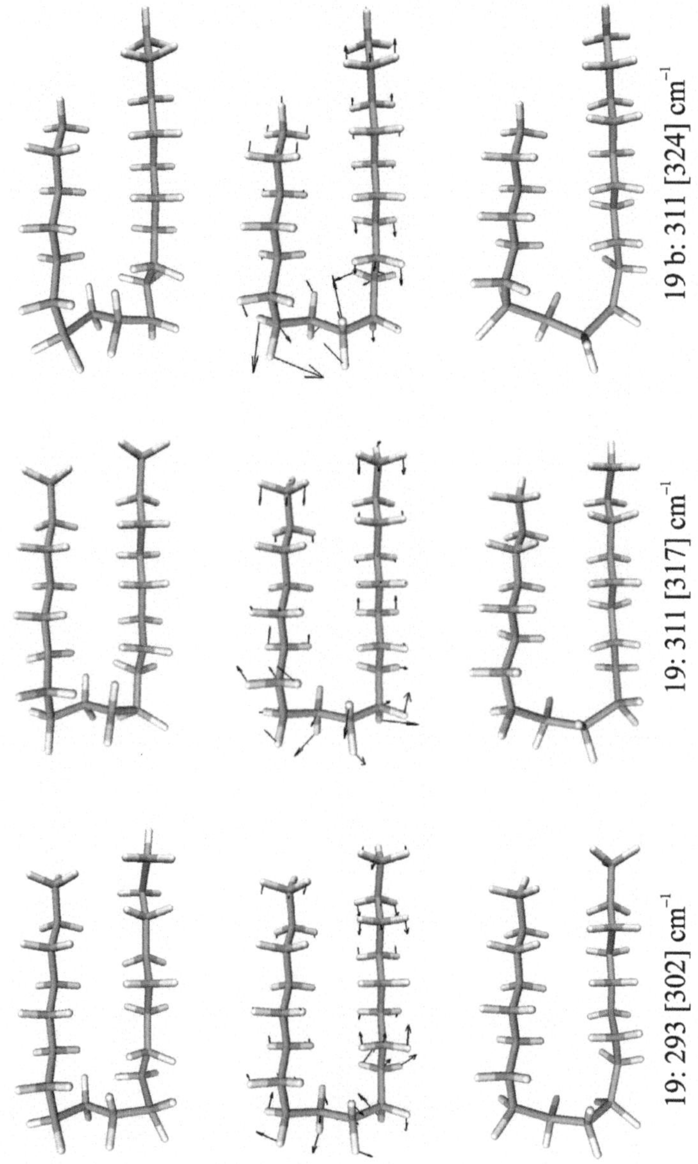

Fig. A.9 Skeletal hairpin vibrations of type IV

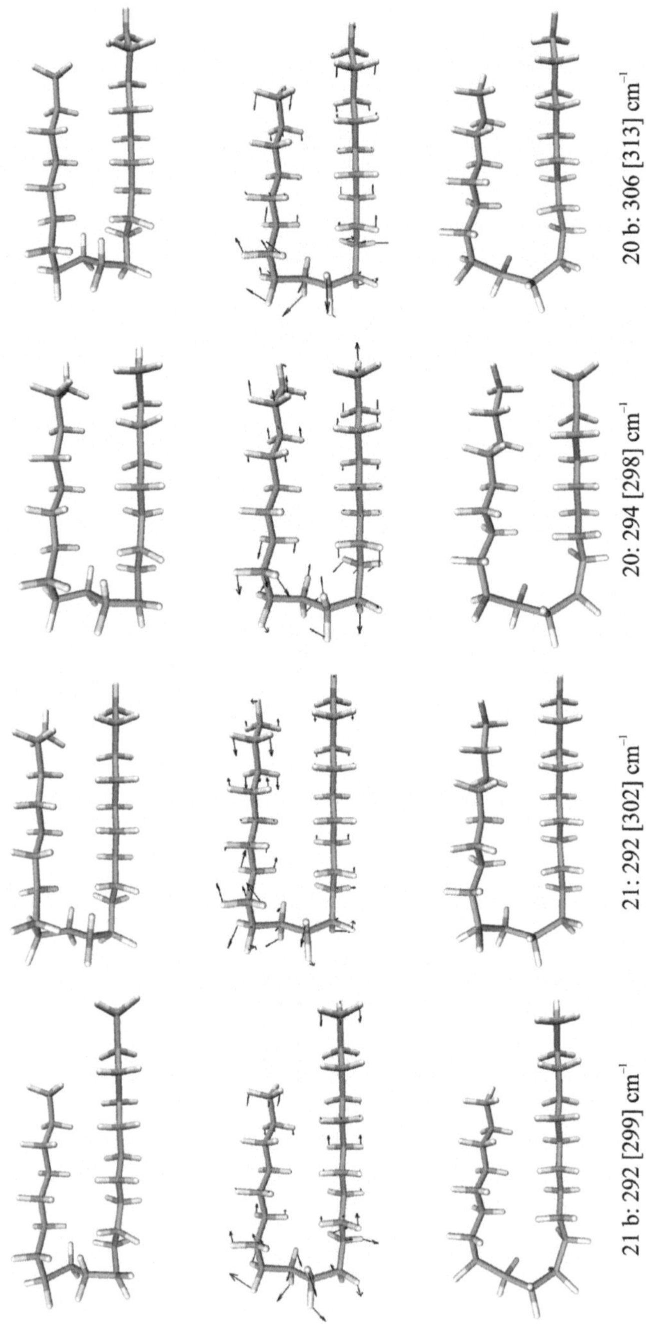

Fig. A.10 Skeletal hairpin vibrations of type IV

A.6 Accordion/CH Ratio

The accordion/CH ratio was derived from Raman jet-measurements of alkanes expanded in helium, helium/argon (6 %), and helium/tetrafluoromethane (4 %), using the spectroscopy software Opus v6.5[2] to integrate spectra of the CH-stretching band and to fit spectra of the accordion vibration band. CH-stretching spectra and spectra of the accordion vibration band were measured in alternation to attenuate fluctuations of substance concentration. The CH-stretching band was integrated in the wavenumber interval 2810–3020 cm^{-1} throughout, while the accordion vibration band and neighboring bands were fitted with Lorentzian curves.

States perturbing the accordion vibration (see Sect. 6.1) were identified by relaxation experiments with carrier gas additives (relative intensity to accordion vibration band does not change) and quantum chemical calculations on the B3LYP/6-311++G** level performed with Turbomole v6.3 [11], as well as calculations on the B3LYP/6-311++G(d,p) level with Gaussian 09 [12]. Quantum chemical calculations were used to identify fundamentals mixing with the accordion vibration and to estimate the occurrence of two-quanta excitations with energies close to the accordion vibration wavenumber and appropriate symmetry to provoke a Fermi resonance. Fitting models of the accordion vibration and neighboring signals are shown in Fig. A.11. Some of the designated perturbations might in fact be bands of single gauche conformers. At the time these fits were made, calculations of single gauche conformer spectra were not yet available. Revision of the fitting models considering single gauche conformers would be possible, but not worthwhile anymore.

Calculated accordion/CH ratios are summarized in Table A.1. This data set is a revision of the Turbomole v6.3 calculations mentioned above, calculated with Turbomole v6.4 including Grimme's D3 correction [8]. Ratios are calculated from scattering cross-sections weighted according to the polarization dependent sensitivity of the curry-jet spectrograph (see Sect. 2.1). The table includes values calculated

Table A.1 Accordion/CH ratio in percent, calculated on the B3LYP-D3/6-311++G** level with Turbomole v6.4 [1] for different vibrational temperatures (T_{vib})

T_{vib}/K	$n = 13$	14	15	16	17	18	19	20
Considering perturbation								
50	9.15	8.98	8.80	8.59	8.41	8.30	8.01	7.91
100	9.89	9.84	9.79	9.69	9.63	9.63	9.42	9.43
150	11.19	11.25	11.33	11.32	11.37	11.46	11.32	11.41
Neglecting perturbation								
50	8.08	8.92	8.57	8.49	8.31	8.00	7.95	4.04
100	8.72	9.79	9.53	9.58	9.51	9.29	9.35	4.79
150	9.84	11.20	11.02	11.20	11.22	11.07	11.22	5.78

[2] http://www.bruker.com/products/optical-spectroscopy/opus-spectroscopy-software/overview.html (23 September 2013).

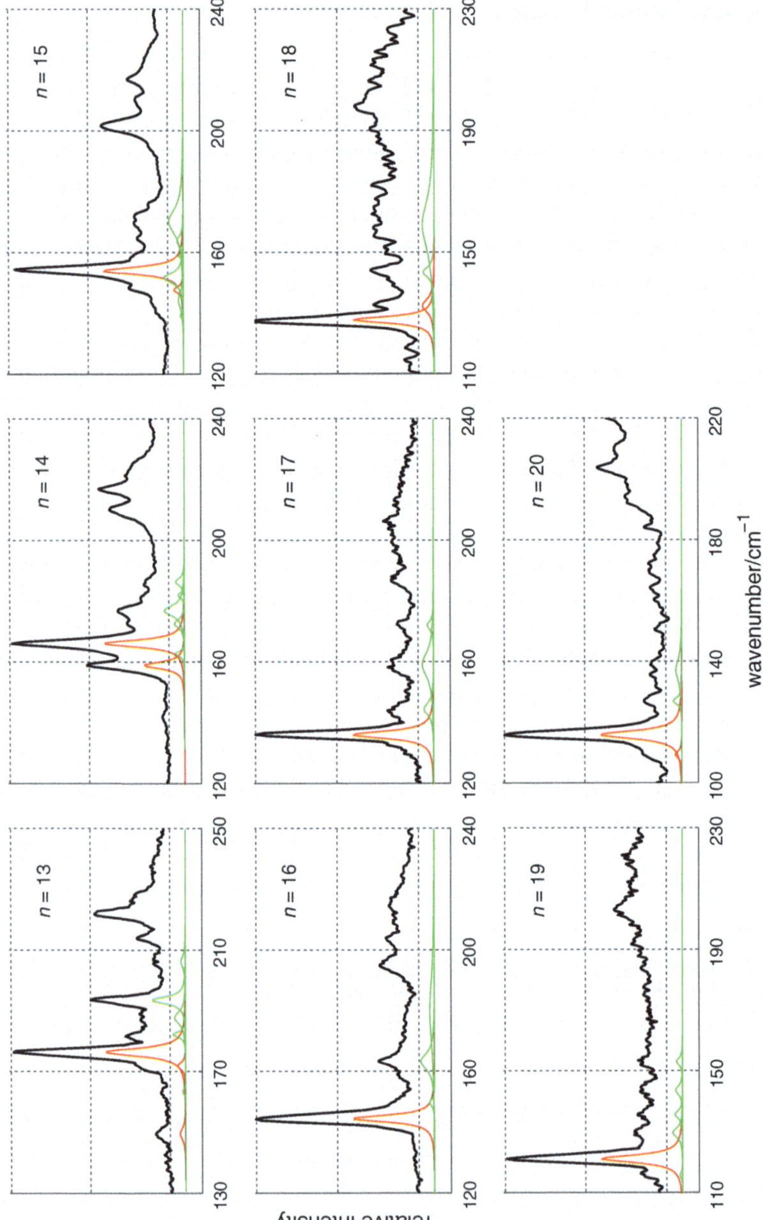

Fig. A.11 Lorentzian fitting models. The accordion vibration band and attendant perturbations are drawn in *red*. Note that the intensities of the Lorentzian curves are scaled to half the accordion vibration band intensity and that spectra are from an early stage of the alkane-folding study

considering harmonic mode mixing of the accordion vibration with the nearest totally symmetric vibration, and such omitting any perturbation. The deviation of these values is significant only in case of tridecane and eicosane. The slight chain length dependence of the calculated accordion/CH ratio was omitted, leading to a predicted ratio of $\approx$10 % at 100 K (quoted as 0.1 in Sect. 4.5). Allowing for a variation of the unknown vibrational temperature between 50 and 150 K leads to an uncertainty of about $\pm$1 %.

A.7 Dimensions of Extended (Perfluoro-)Alkanes

See Figs. A.12, A.13 and A.14.

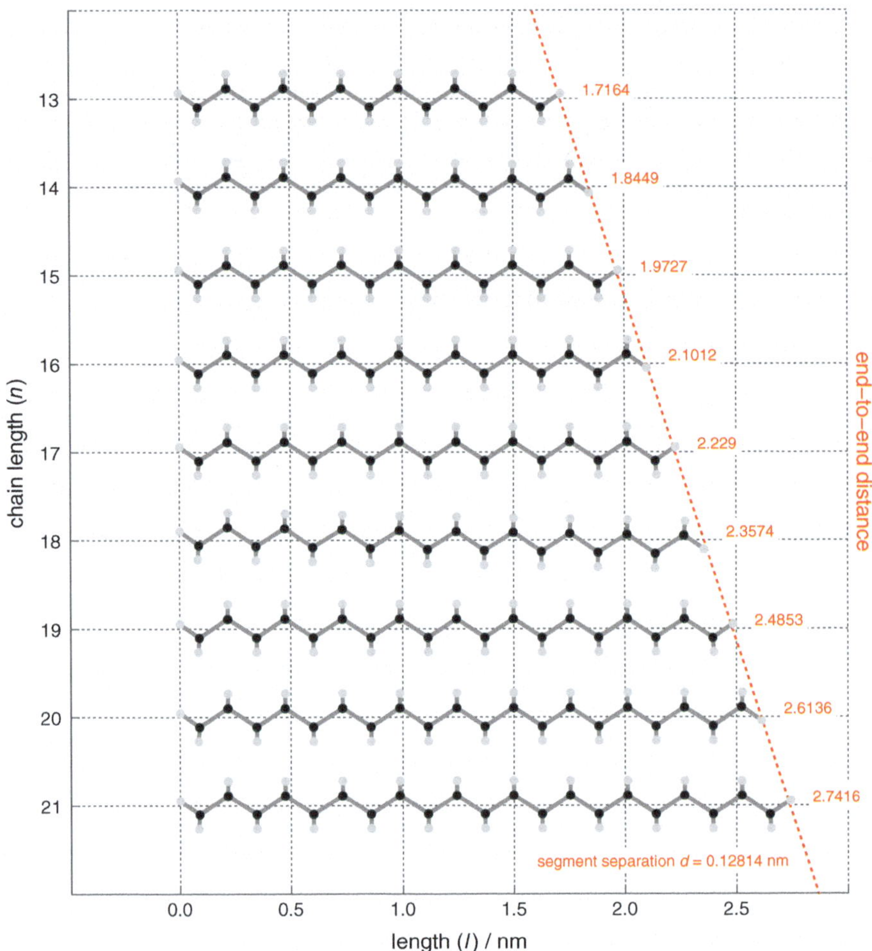

Fig. A.12 Projection of all-trans alkane atom coordinates on the plane spanned by carbon atoms (optimized structures on the B3LYP-D3/6-311++G** level, Turbomole v6.4 [1]). The end-to-end distance, given on the *right side*, yields a methylene separation of 128 pm in the l-direction. Reference II—Reproduced by permission of The Royal Society of Chemistry

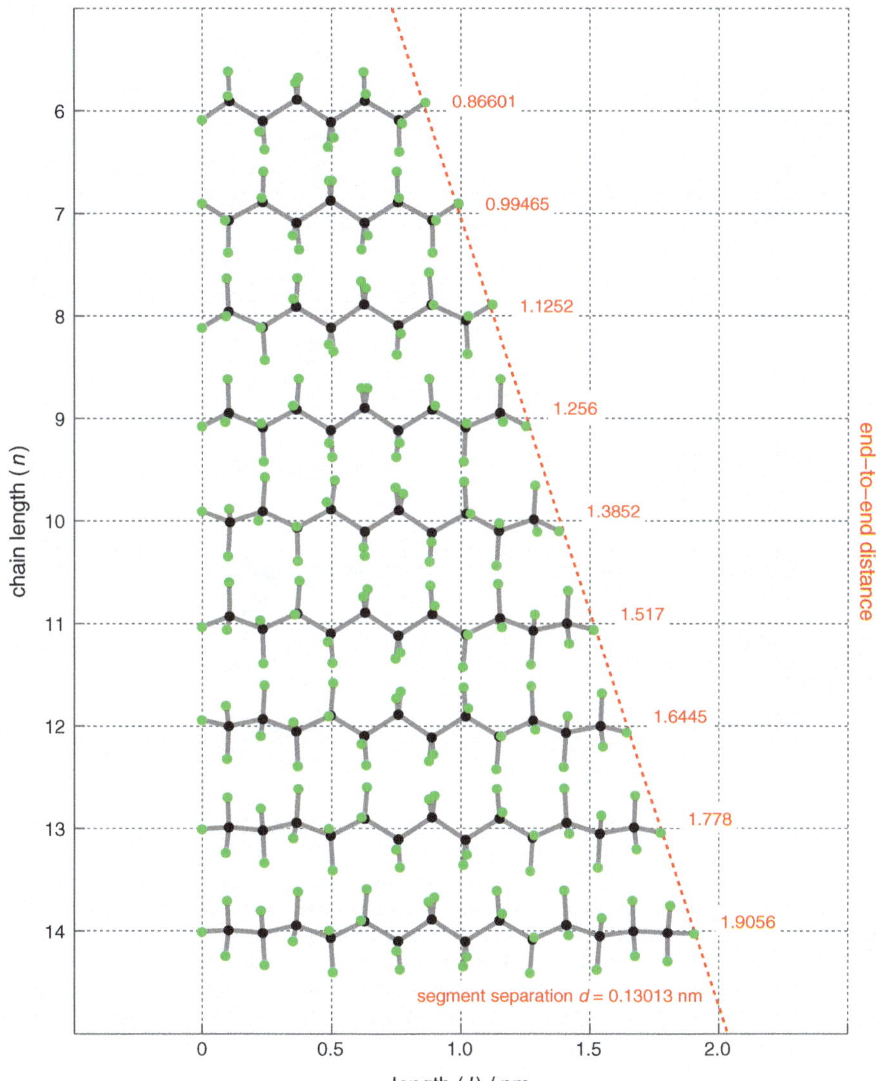

Fig. A.13 Projection of all-anti perfluoroalkane atom coordinates (optimized structures on the B3LYP-D3/def2-TZVP level, Turbomole v6.4 [1]). The end-to-end distance, given on the *right side*, yields a CF_2 separation of 130 pm in the l-direction

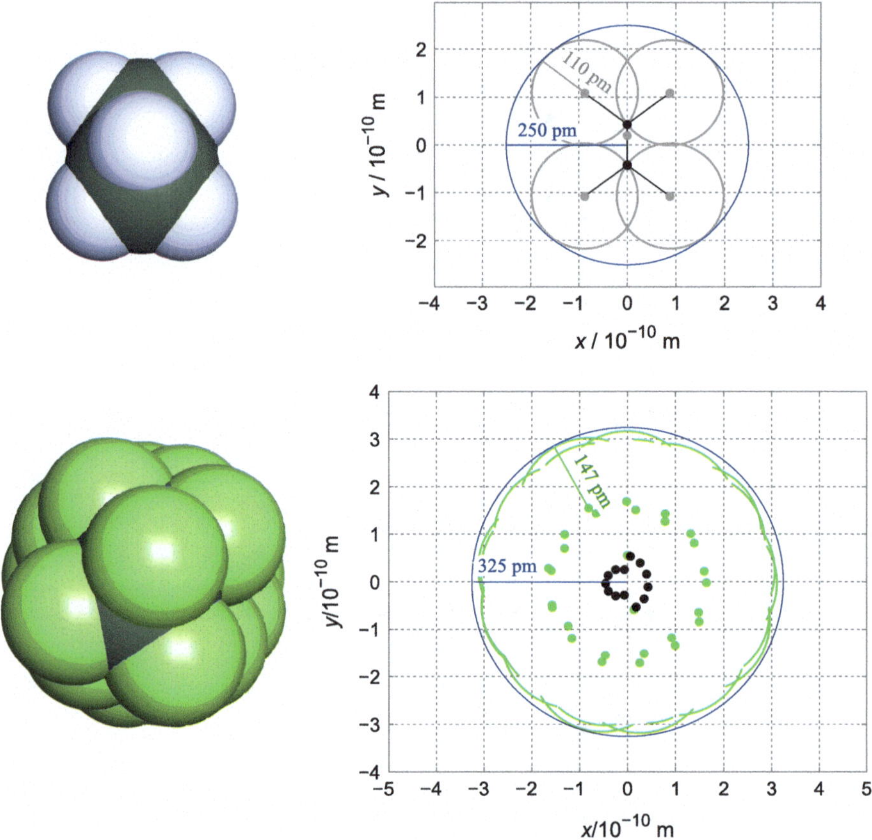

Fig. A.14 Projection of all-trans tridecane and all-anti perfluorotridecane atom coordinates on the plane perpendicular to the long axes of the molecules (optimized structures on the B3LYP-D3/6-311++G** and B3LYP-D3/def2-TZVP level, respectively, Turbomole v6.4 [1]). In case of perfluorotridecane, bonds are omitted and fluorine van-der-Waals spheres are restricted to the outer boundary for better clarity. The spatial occupation is approximated by a cylinder with a radius set such that all hydrogen and fluorine atoms fit according to their van-der-Waals radii (110 and 147 pm, respectively [13, 14])

A.8 C–C–C Bending Frequency Branch

From harmonic frequency calculations on the B3LYP-D3/6-311++G** level used to simulate Raman spectra, the polyethylene frequency branches can be approximated. This was done for the C–C–C bending frequency branch to which the longitudinal acoustic modes belong, in order to elucidate harmonic mode mixing of LAM-1 modes. The frequency branch can actually be calculated for an infinite all-trans chain using periodic internal coordinates [15]. Here, the frequency branch is not directly

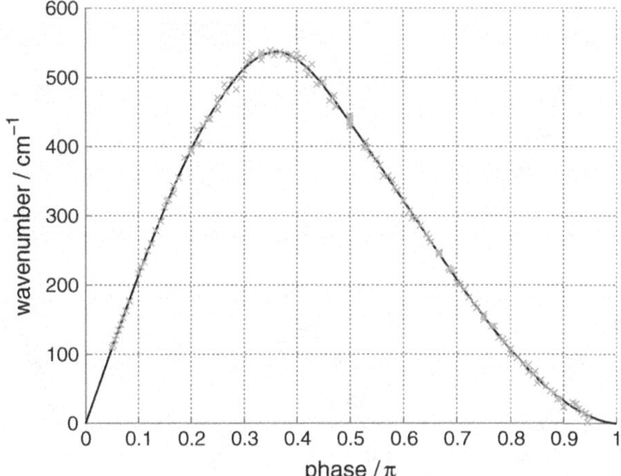

Fig. A.15 C–C–C bending frequency branch from B3LYP-D3/6-311++G** harmonic frequency calculations (n = 13–21). Calculated wavenumbers are drawn with ×, the interpolation with a *solid line*

calculated, but interpolated[3] from the available quantum chemical calculations of vibrations of finite chains with n = 13–21. Finite chain vibrations will scatter somewhat around the frequency branch of the infinite system, due to end-group effects and mode mixing. Interpolating the full frequency branch from this data can thus only result in an approximation to the "real" frequency branch which would result from a calculation on the equivalent level on the infinite system. However, the resulting branch correctly predicts the occurrence of LAM-1 mode mixing at certain chain lengths and is thus sufficiently accurate for this work. It is depicted in Fig. A.15. The involved calculated normal modes were assigned to k values considering the known form of the C–C–C bending frequency branch (vibrational wavenumbers first increase, reach a maximum, and then decrease with increasing k) and symmetry requirements (A_1/A_g symmetry for odd k and B_1/B_u symmetry for even k). The phase is calculated *via* $\varphi = k\pi/(n-1)$, $k = 1, 2, \ldots, n-2$ [17]. Note that Snyder [17] gives a relation involving the number of coupled oscillators ($m = n - 2$ in this case of coupled C–C–C bending oscillators).

Figure A.16 shows how unperturbed C–C–C bending modes are distributed over the frequency branch for different chain lengths, see Fig. 6.2 and discussion in Chap. 6.

[3] The interpolant was made by smoothing the data using Matlab [16] (method "loess" = locally weighted scatter plot smooth) and fitting a cubic spline.

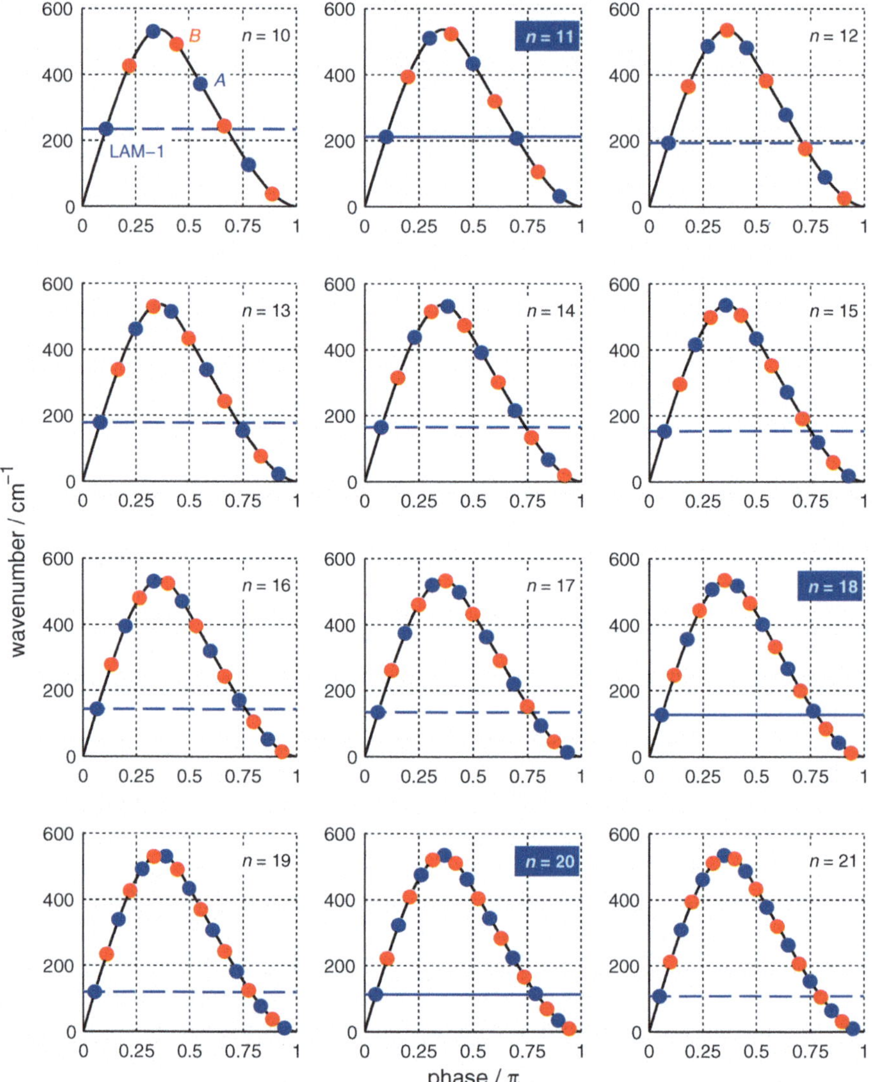

Fig. A.16 Distribution of modes over the C–C–C bending frequency branch for $n = 13$–21. *Colors* indicate the symmetry of the modes: *blue* = $A_{1(g)}$, *red* = $B_{2(u)}$. The LAM-1 is the totally symmetric (*blue*) mode with the lowest phase value. Perturbation of the LAM-1 occurs when its frequency comes close to the frequency of a totally symmetric mode at the high-phase value end of the branch ($n = 11, 18, 20$). The LAM-1 frequency is highlighted with a *blue line*

A.9 Subsidiary Figures

See Figs. A.17–A.24.

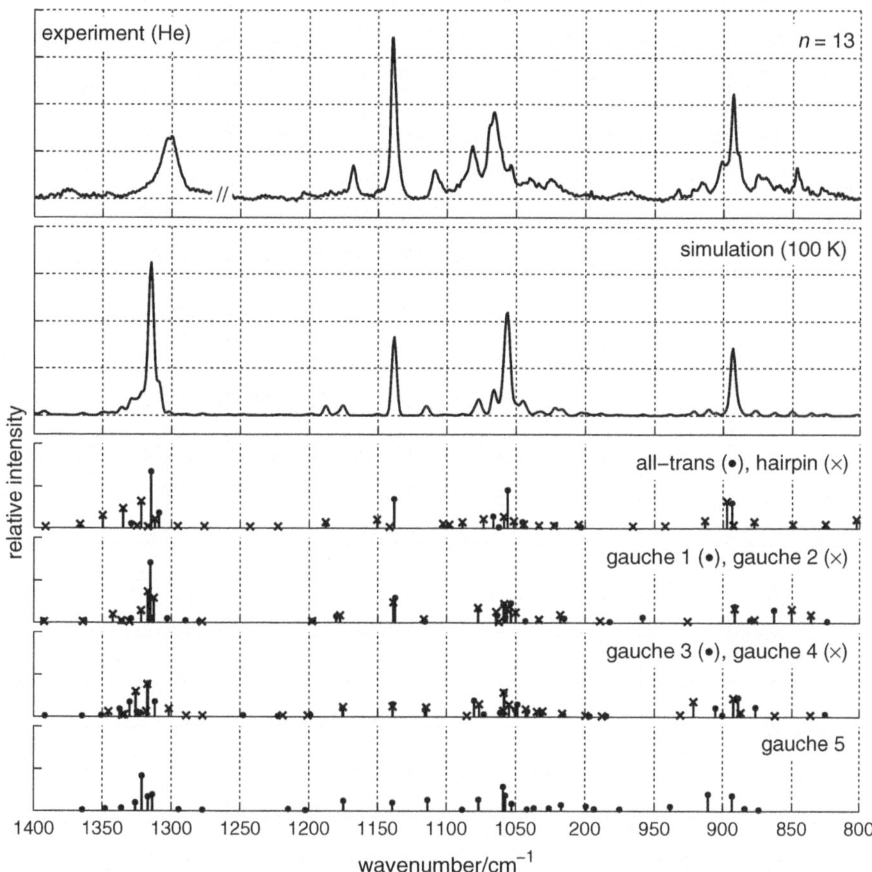

Fig. A.17 Comparison of an averaged jet-cooled Raman spectrum of tridecane to a B3LYP-D3/6-311++G** simulation in the C–C stretching region. The experimental spectrum is Savitzky-Golay filtered (7 pt.). *Measurement conditions* expansion in He, $\vartheta_s = 35\,°C$, $p_0 = 0.5$ bar, $p_b = 0.9$ mbar, $d_n = 1$ mm, exposure 6×600 s. *Calculation* effective temperature used in single gauche to all-trans weighting = 100 K, hairpin to all-trans weighting = 0.1:1, wavenumber scaling factor = 0.99

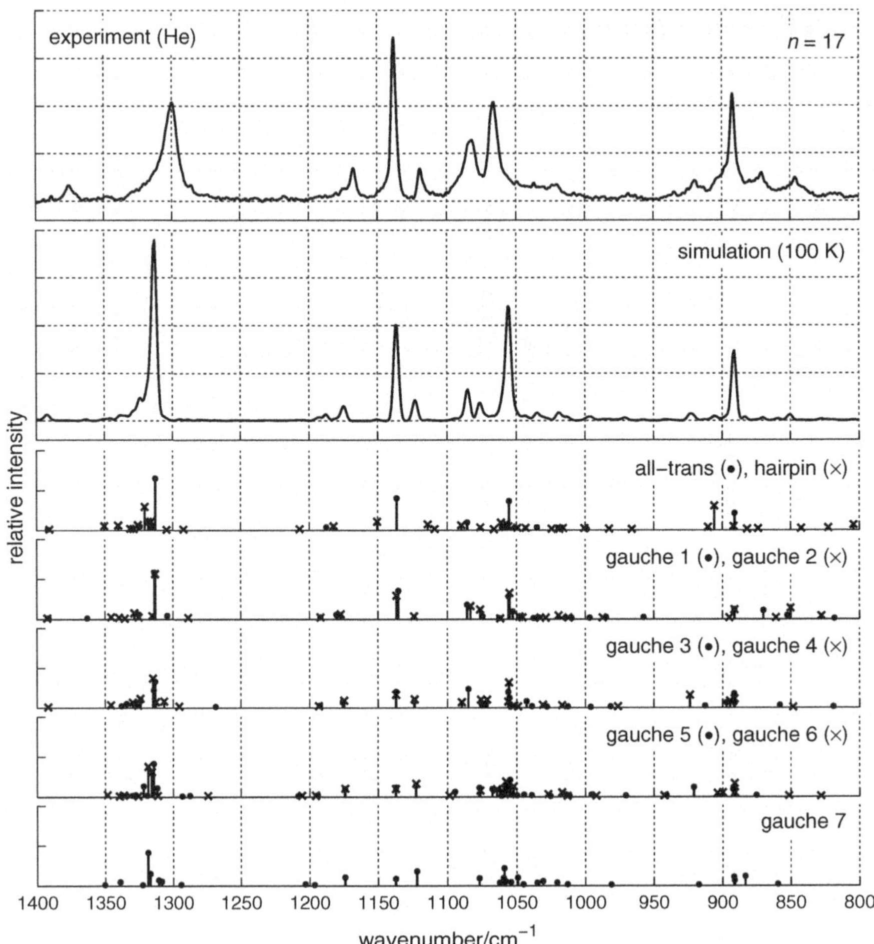

Fig. A.18 Comparison of an averaged jet-cooled Raman spectrum of heptadecane to a B3LYP-D3/6-311++G** simulation in the C–C stretching region. The experimental spectrum is Savitzky-Golay filtered (7 pt.). *Measurement conditions* expansion in He, $\vartheta_s = 75°C$, $p_0 = 0.5$ bar, $p_b = 0.8$ mbar, $d_n = 1$ mm, exposure 6×600 s. *Calculation* effective temperature used in single gauche to all-trans weighting = 100 K, hairpin to all-trans weighting =0.1:1, wavenumber scaling factor = 0.99

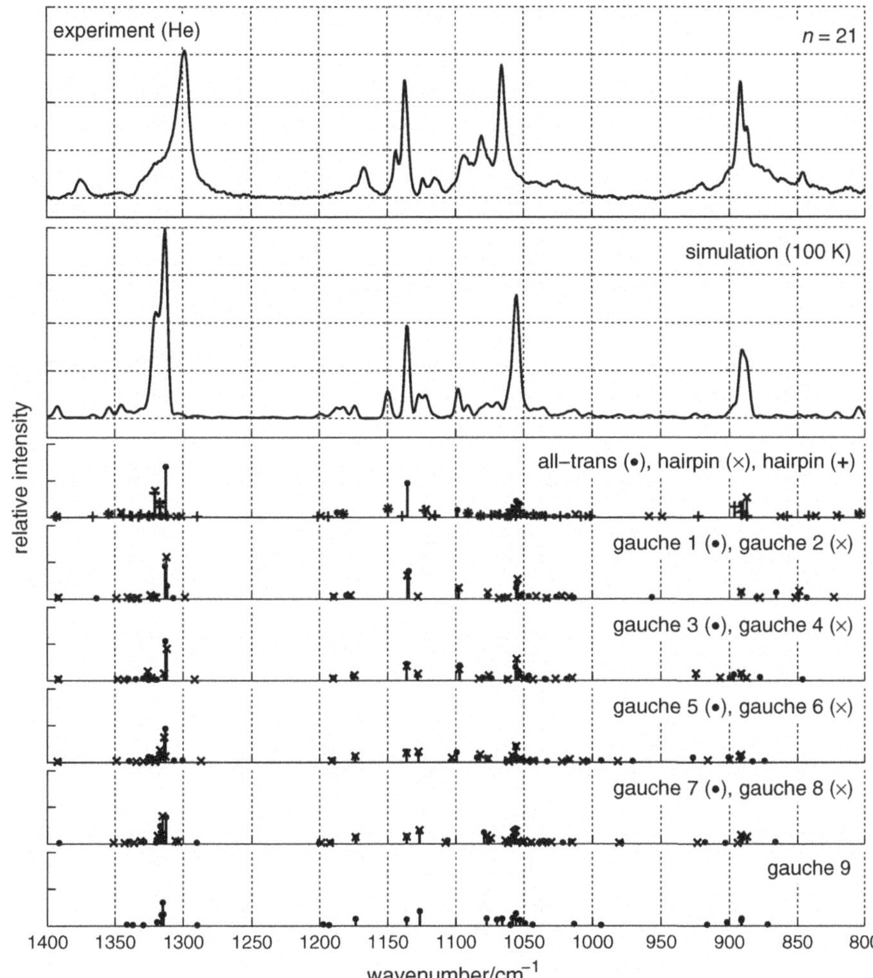

Fig. A.19 Comparison of an averaged jet-cooled Raman spectrum of heneicosane to a B3LYP-D3/6-311++G** simulation in the C–C stretching region. The experimental spectrum is Savitzky-Golay filtered (7 pt.). *Measurement conditions* expansion in He, $\vartheta_s = 125°C$, $p_0 = 0.55$ bar, $p_b = 0.8$ mbar, $d_n = 1$ mm, exposure 6×600 s. *Calculation* effective temperature used in single gauche to all-trans weighting = 100 K, hairpin (hairpin b) to all-trans weighting = 1.7 (0.6):1, wavenumber scaling factor = 0.99

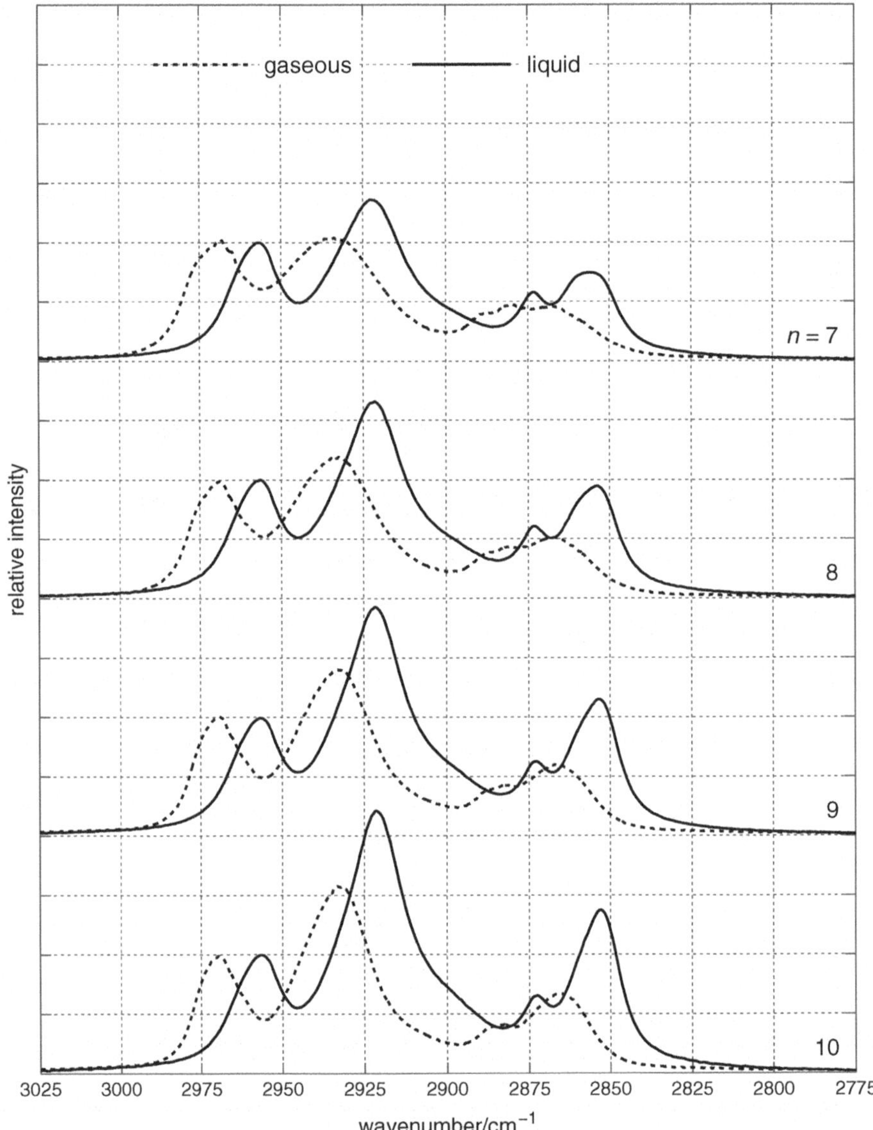

Fig. A.20 Comparison of FTIR gas phase and liquid phase spectra of short alkanes at room temperature in the C–H stretching region. Liquid phase spectra were measured with a Vector 22 spectrometer (Bruker) at 1 cm^{-1} resolution using a ZnSe ATR cell, vapor spectra (1 mbar alkane mixed with 500 mbar N_2) were measured with a Vertex 70v spectrometer (Bruker) at 0.5 cm^{-1} resolution using a 23 cm long sample cell equipped with KBr windows

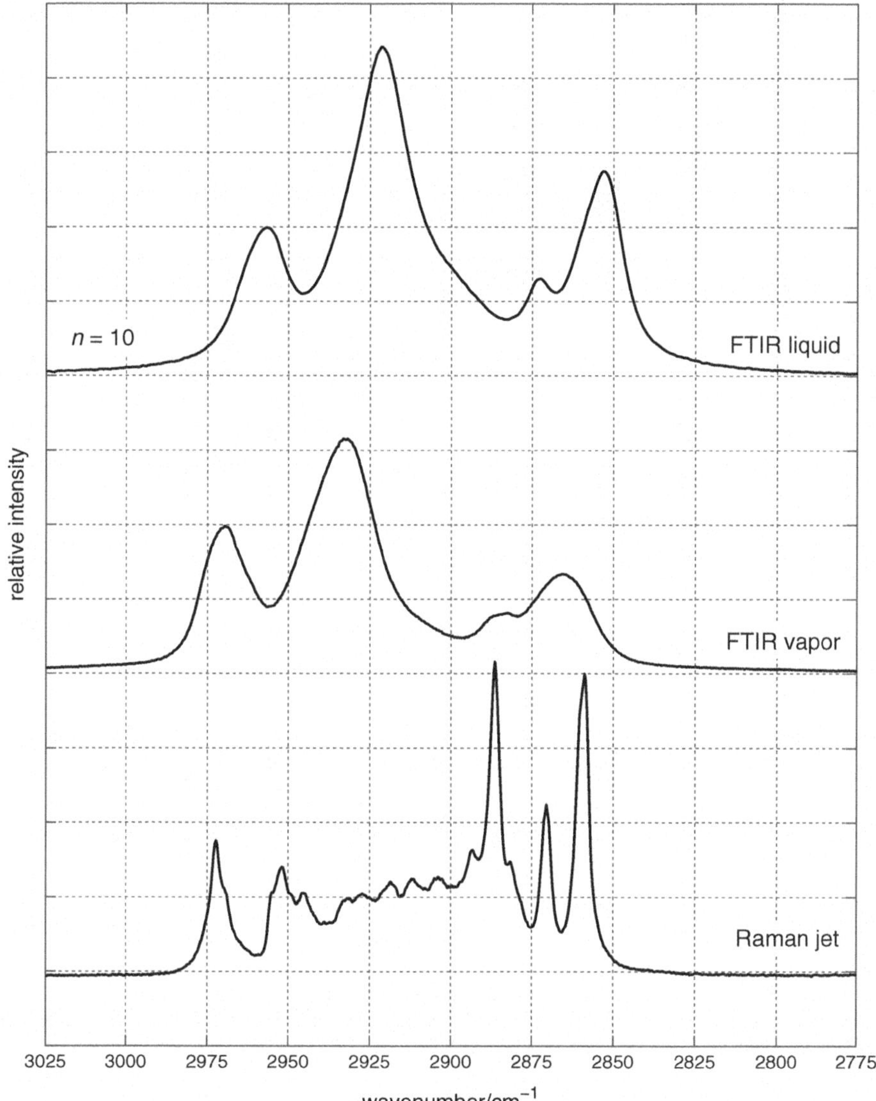

Fig. A.21 Comparison of *n*-decane spectra in the C–H stretching region under different physical conditions. The FTIR measurement conditions are described in the caption of Fig. A.20, Raman jet measurement conditions were as follows: expansion in He, $\vartheta_s = 12\,°C$, $p_0 = 0.5$ bar, $p_b = 0.9$ mbar, $d_n = 1$ mm, exposure 6×90 s. Helium/decane mixtures were prepared with the glass saturator flushed with helium at 1.9 bar

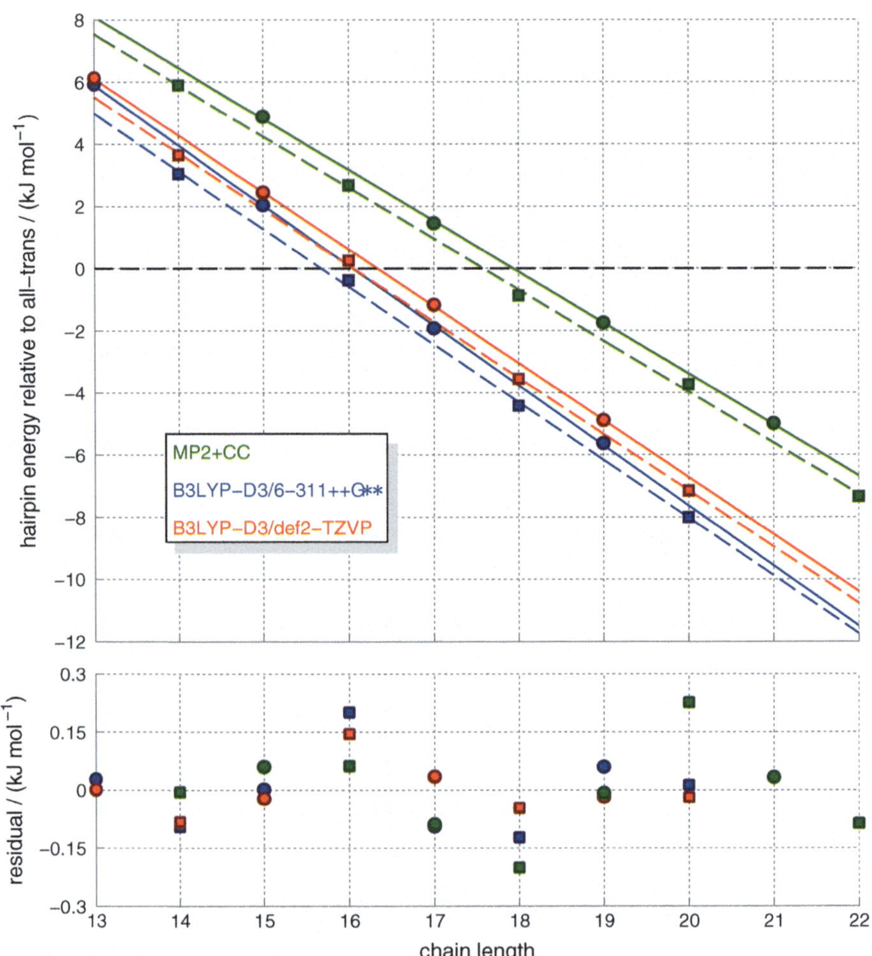

Fig. A.22 Hairpin energy relative to all-trans energy (electronic energy + zero-point vibrational energy) on different computational levels, see Sect. 4.5.2. *Lines* are drawn to guide the eye and to demonstrate linearity (*dashed* even-numbered chain length, *solid* odd-numbered chain length)

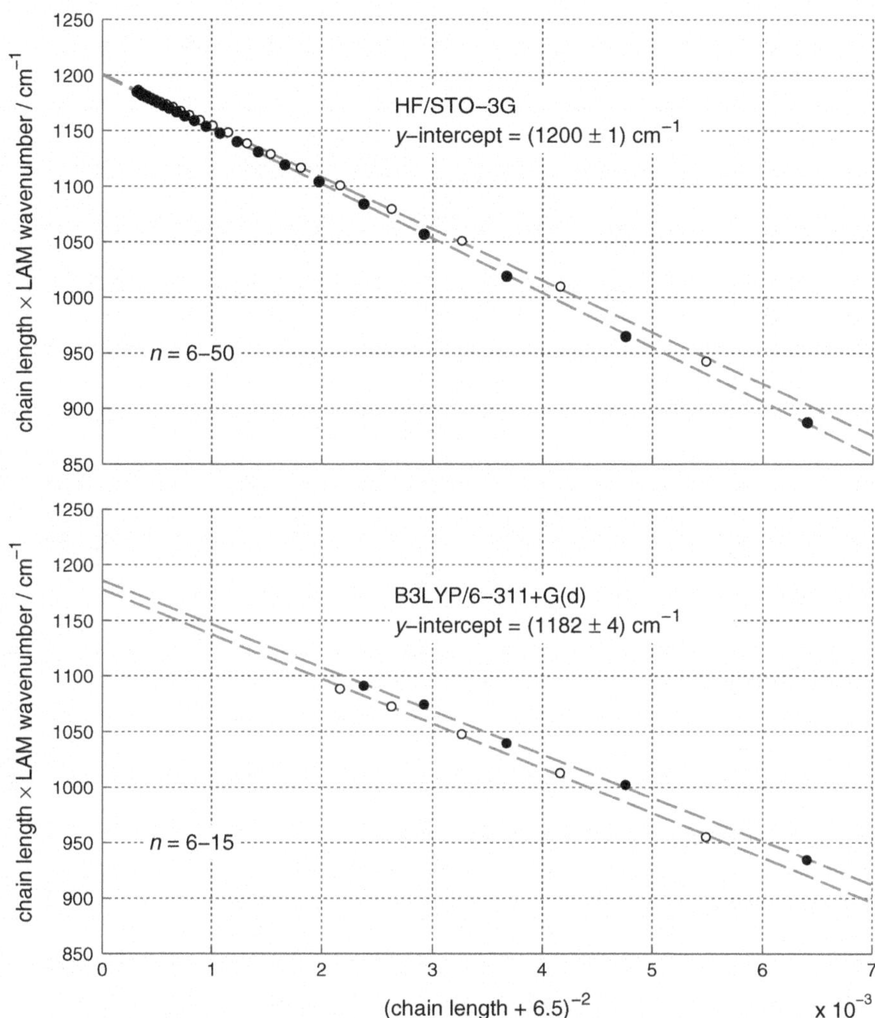

Fig. A.23 LAM-1 frequency extrapolation from calculations of perfluorinated alkanes in all-anti conformation, see Sect. 6.2.2. Calculations were done using Gaussian 03 [18, 19]. *Filled symbols* indicate even-numbered chain lengths, *open symbols* odd-numbered chain lengths. LAM-1 frequencies are deperturbed using Eq. 6.4, considering only the totally symmetric TAM with the closest wavenumber. Reproduced and adapted material from reference III

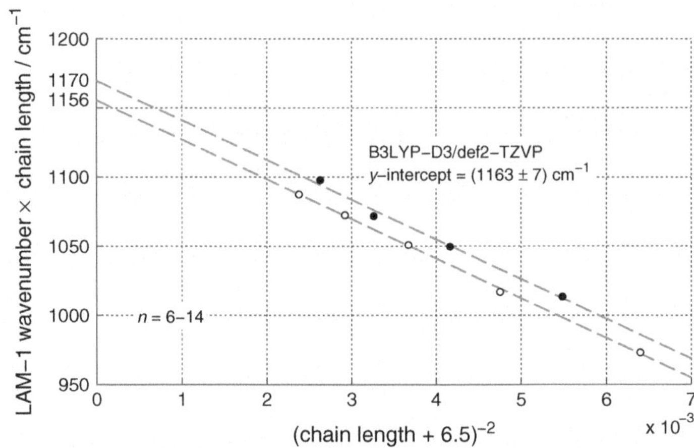

Fig. A.24 LAM-1 frequency extrapolation from calculations of perfluorinated alkanes in all-anti conformation, see Sect. 6.2.2. Calculations were done using Turbomole v6.4 [1]. *Filled symbols* indicate even-numbered chain lengths, *open symbols* odd-numbered chain lengths. LAM-1 frequencies are deperturbed using Eq. 6.4, considering only the totally symmetric TAM with the closest wavenumber

A.10 Chemicals

See Table A.2.

Table A.2 List of used chemicals (not further purified)

Compound		CAS	Purity/%	Supplier	Comment[a]
Perfluoroalkanes					
Perfluorohexane	C_6F_{14}	355-42-0	99	Sigma-Aldrich	PN 281042 LOT MKBD0617
Perfluorooctane	C_8F_{18}	307-34-6	98	Sigma-Aldrich	PN 359238
Perfluorononane	C_9F_{20}	375-96-2	99	Fluorochem	PN 007118 BN 2737
Perfluorodecane	$C_{10}F_{22}$	307-45-9	97	Apollo-Scientific	PN PC2903 BN AS413408
Perfluorododecane	$C_{12}F_{26}$	307-59-5	97	Fluorochem	PN 007036
Perfluorotridecane	$C_{13}F_{28}$	376-03-4	98	Fluorochem	PN 009331 BN 1458
Perfluorotetradecane	$C_{14}F_{30}$	307-62-0	97	Apollo-Scientific	PN PC6216 BN AS413982

(continued)

Table A.2 (continued)

Compound		CAS	Purity/%	Supplier	Comment[a]
Gases					
Helium		7440-59-7	99.996	Linde	
Argon		7440-37-1	99.999	Air Liquide	
Tetrafluoromethane	CF_4	75-73-0	99.999	Linde	
Hexafluoroethane	C_2F_6	76-16-4	99.999	Linde	
Sulfur hexafluoride	SF_6	2551-62-4	99.9	Linde	
Helium/neon (30 %)				Linde	Prüfgasklasse 1
Alkanes					
n-pentane	C_5H_{12}	109-66-0	99	Sigma-Aldrich	PN 34956 LOT SZBA170M
n-hexane	C_6H_{14}	110-54-3	99	Merck	Measurement from Ref. [20]
n-decane	$C_{10}H_{22}$	124-18-5	99	Sigma-Aldrich	PN D901 LOT MKBG3224V
n-undecane	$C_{11}H_{24}$	1120-21-4	99	Sigma-Aldrich	PN U407 LOT 06010EE-337
n-dodecane	$C_{12}H_{26}$	112-40-3	99	Sigma-Aldrich	PN D221104 LOT 08305CH-317
n-tridecane	$C_{13}H_{28}$	629-50-5	99	Sigma-Aldrich	PN T57401 LOT 04220LE-267
n-tetradecane	$C_{14}H_{30}$	629-59-4	99	Sigma-Aldrich	PN 87140 LOT BCBC1627
n-pentadecane	$C_{15}H_{32}$	629-62-9	99	Sigma-Aldrich	PN P3406 LOT 04414HH-118
n-hexadecane	$C_{16}H_{34}$	544-76-3	99	Alfa Aesar	PN A10322 LOT 10141578
n-heptadecane	$C_{17}H_{36}$	629-78-7	98	Sigma-Aldrich	PN 51580 LOT 1350580
n-octadecane	$C_{18}H_{38}$	593-45-3	99	Alfa Aesar	PN 31954 LOT E18W031
n-nonadecane	$C_{19}H_{40}$	629-92-5	99	Sigma-Aldrich	PN 74160 LOT 440216/1
n-eicosane	$C_{20}H_{42}$	112-95-8	99	Sigma-Aldrich	PN 219274 LOT 04225AJ
n-heneicosane	$C_{21}H_{44}$	629-94-7	99	ABCR	PN AB116964 LOT 1035515

[a] *PN* ≡ product number/item number, *LOT* ≡ lot number, *BN* ≡ batch number

References

1. TURBOMOLE v6.4 2012, A development of University of Karlsruhe and Forschungszentrum Karlsruhe GmbH, 1989–2007, TURBOMOLE GmbH, since 2007; available from http://www.turbomole.com
2. C. Hättig, F. Weigend, CC2 excitation energy calculations on large molecules using the resolution of the identity approximation. J. Chem. Phys. **113**, 5154–5161 (2000)
3. D. Gruzman, A. Karton, J.M.L. Martin, Performance of ab initio and density functional methods for conformational equilibria of C_nH_{2n+2} alkane isomers ($n = 4$–8). J. Phys. Chem. A **113**, 11974–11983 (2009)
4. S. Tsuzuki, L. Schäfer, H. Gotō, E.D. Jemmis, H. Hosoya, K. Siam, K. Tanabe, E. Ōsawa, Investigation of intramolecular interactions in *n*-alkanes. Cooperative energy increments associated with GG and GTG' sequences. J. Am. Chem. Soc. **113**, 4665–4671 (1991)
5. J. Grant Hill, J.A. Platts, H.-J. Werner, Calculation of intermolecular interactions in the benzene dimer using coupled-cluster and local electron correlation methods. Phys. Chem. Chem. Phys. **8**, 4072–4078 (2006)
6. K.E. Riley, P. Hobza, Assessment of the MP2 method, along with several basis sets, for the computation of interaction energies of biologically relevant hydrogen bonded and dispersion bound complexes. J. Phys. Chem. A **111**, 8257–8263 (2007)
7. R.A. Bachorz, F.A. Bischoff, S. Höfener, W. Klopper, P. Ottiger, R. Leist, J.A. Frey, S. Leutwyler, Scope and limitations of the SCS-MP2 method for stacking and hydrogen bonding interactions. Phys. Chem. Chem. Phys. **10**, 2758–2766 (2008)
8. S. Grimme, J. Antony, S. Ehrlich, H. Krieg, A consistent and accurate ab initio parametrization of density functional dispersion correction (DFT-D) for the 94 elements H-Pu. J. Chem. Phys. **132**, 154104 (2010)
9. M.R. Munrow, R. Subramanian, A.J. Minei, D. Antic, M.K. MacLeod, J. Michl, R. Crespo, M.C. Piqueras, M. Izuha, T. Ito, Y. Tatamitani, K. Yamanou, T. Ogata, S.E. Novick, Rotational spectra of gauche perfluoro-*n*-butane, C_4F_{10}; perfluoro-iso-butane, $(CF_3)_3CF$; and tris(trifluoromethyl)methane, $(CF_3)_3CH$. J. Mol. Spectrosc. **242**, 129–138 (2007)
10. C.J. Cramer, *Essentials of Computational Chemistry: Theories and Models*, 2nd edn. (Wiley, Chichester, 2004)
11. TURBOMOLE v6.3 2011, A development of University of Karlsruhe and Forschungszentrum Karlsruhe GmbH, 1989–2007, TURBOMOLE GmbH, since 2007; available from http://www.turbomole.com
12. M.J. Frisch, G.W. Trucks, H.B. Schlegel, G.E. Scuseria, M.A. Robb, J.R. Cheeseman, G. Scalmani, V. Barone, B. Mennucci, G.A. Petersson, H. Nakatsuji, M. Caricato, X. Li, H.P. Hratchian, A.F. Izmaylov, J. Bloino, G. Zheng, J.L. Sonnenberg, M. Hada, M. Ehara, K. Toyota, R. Fukuda, J. Hasegawa, M. Ishida, T. Nakajima, Y. Honda, O. Kitao, H. Nakai, T. Vreven, J.A. Montgomery, Jr., J.E. Peralta, F. Ogliaro, M. Bearpark, J.J. Heyd, E. Brothers, K.N. Kudin, V.N. Staroverov, R. Kobayashi, J. Normand, K. Raghavachari, A. Rendell, J.C. Burant, S.S. Iyengar, J. Tomasi, M. Cossi, N. Rega, J.M. Millam, M. Klene, J.E. Knox, J.B. Cross, V. Bakken, C. Adamo, J. Jaramillo, R. Gomperts, R.E. Stratmann, O. Yazyev, A.J. Austin, R. Cammi, C. Pomelli, J.W. Ochterski, R.L. Martin, K. Morokuma, V.G. Zakrzewski, G.A. Voth, P. Salvador, J.J. Dannenberg, S. Dapprich, A.D. Daniels, O. Farkas, J.B. Foresman, J.V. Ortiz, J. Cioslowski, D.J. Fox, Gaussian 09, Revision A.02. Gaussian Inc, Wallingford (2009)
13. M. Mantina, A.C. Chamberlin, R. Valero, C.J. Cramer, D.G. Truhlar, Consistent van der Waals radii for the whole main group. J. Phys. Chem. A **113**, 5806–5812 (2009)
14. A. Bondi, van der Waals volumes and radii. J. Phys. Chem. **68**, 441–451 (1964)
15. M. Tasumi, T. Shimanouchi, T. Miyazawa, Normal vibrations and force constants of polymethylene chain. J. Mol. Spectrosc. **9**, 261–287 (1962)
16. MATLAB, version 7.12.0 (R2011a), The MathWorks Inc., Natick (2011)
17. R.G. Snyder, J.H. Schachtschneider, Vibrational analysis of the *n*-paraffins–I: assignments of infrared bands in the spectra of C_3H_8 through *n*-$C_{19}H_{40}$. Spectrochimica Acta **19**, 85–116 (1963)

18. M.J. Frisch, G.W. Trucks, H.B. Schlegel, G.E. Scuseria, M.A. Robb, J.R. Cheeseman, J.A. Montgomery, Jr., T. Vreven, K.N. Kudin, J.C. Burant, J.M. Millam, S.S. Iyengar, J. Tomasi, V. Barone, B. Mennucci, M. Cossi, G. Scalmani, N. Rega, G.A. Petersson, H. Nakatsuji, M. Hada, M. Ehara, K. Toyota, R. Fukuda, J. Hasegawa, M. Ishida, T. Nakajima, Y. Honda, O. Kitao, H. Nakai, M. Klene, X. Li, J.E. Knox, H.P. Hratchian, J.B. Cross, V. Bakken, C. Adamo, J. Jaramillo, R. Gomperts, R.E. Stratmann, O. Yazyev, A.J. Austin, R. Cammi, C. Pomelli, J.W. Ochterski, P.Y. Ayala, K. Morokuma, G.A. Voth, P. Salvador, J.J. Dannenberg, V.G. Zakrzewski, S. Dapprich, A.D. Daniels, M.C. Strain, O. Farkas, D.K. Malick, A.D. Rabuck, K. Raghavachari, J.B. Foresman, J.V. Ortiz, Q. Cui, A.G. Baboul, S. Clifford, J. Cioslowski, B.B. Stefanov, G. Liu, A. Liashenko, P. Piskorz, I. Komaromi, R.L. Martin, D.J. Fox, T. Keith, M.A. Al-Laham, C.Y. Peng, A. Nanayakkara, M. Challacombe, P.M. W. Gill, B. Johnson, W. Chen, M.W. Wong, C. Gonzalez, J.A. Pople, Gaussian 03, Revision B.04, Gaussian Inc, Wallingford (2004)
19. P. Drawe, Mechanical properties of chain molecules from spectroscopic data, B.Sc. Thesis, Georg-August-Universität Göttingen (2010)
20. T.N. Wassermann, Umgebungseinflüsse auf die C-C- und C-O-Torsionsdynamik in Molekülen und Molekülaggregaten: Schwingungsspektroskopie bei tiefen Temperaturen. Ph.D. Thesis, Georg-August-Universität Göttingen (2009)

The manufacturer's authorised representative in the EU is Springer Nature Customer Service Centre GmbH, Europaplatz 3, 69115 Heidelberg, Germany. If you have any concerns regarding our products, please contact ProductSafety@springernature.com

Printed and bound by CPI Group (UK) Ltd, Croydon, CR0 4YY
16/07/2026
02169460-0001